C0-AWL-499

COMPACT Research

Animal Experimentation

by Peggy J. Parks

Current Issues

San Diego, CA

For more information, contact:
ReferencePoint Press, Inc.
PO Box 27779
San Diego, CA 92198
www. ReferencePointPress.com

Picture credits:
Maury Aaseng, 33–36, 50–53, 67–69, 83–85
AP Images, 10, 19

LIBRARY OF CONGRESS CATALOGING-IN-PUBLICATION DATA

Parks, Peggy J., 1951–
Animal experimentation / by Peggy J Parks.
p. cm. — (Compact research)
Includes bibliographical references and index.
ISBN-13: 978-1-60152-037-1 (hardback)
ISBN-10: 1-60152-037-9 (hardback)
1. Animal experimentation—Moral and ethical aspects—United States. 2. Animal rights—United States. 3. Animal welfare—United States. I. Title.
HV4930.P37 2008
179'.4—dc22

2007036725

Contents

Foreword

"Where is the knowledge we have lost in information?"

—"The Rock," T.S. Eliot.

As modern civilization continues to evolve, its ability to create, store, distribute, and access information expands exponentially. The explosion of information from all media continues to increase at a phenomenal rate. By 2020 some experts predict the worldwide information base will double every 73 days. While access to diverse sources of information and perspectives is paramount to any democratic society, information alone cannot help people gain knowledge and understanding. Information must be organized and presented clearly and succinctly in order to be understood. The challenge in the digital age becomes not the creation of information, but how best to sort, organize, enhance, and present information.

ReferencePoint Press developed the *Compact Research* series with this challenge of the information age in mind. More than any other subject area today, researching current events can yield vast, diverse, and unqualified information that can be intimidating and overwhelming for even the most advanced and motivated researcher. The *Compact Research* series offers a compact, relevant, intelligent, and conveniently organized collection of information covering a variety of current and controversial topics ranging from illegal immigration to marijuana.

The series focuses on three types of information: objective single-author narratives, opinion-based primary source quotations, and facts and statistics. The clearly written objective narratives provide context and reliable background information. Primary source quotes are carefully selected and cited, exposing the reader to differing points of view. And facts and statistics sections aid the reader in evaluating perspectives. Presenting these key types of information creates a richer, more balanced learning experience.

For better understanding and convenience, the series enhances information by organizing it into narrower topics and adding design features that make it easy for a reader to identify desired content. For example, in *Compact Research: Illegal Immigration*, a chapter covering the economic impact of illegal immigration has an objective narrative explaining the various ways the economy is impacted, a balanced section of numerous primary source quotes on the topic, followed by facts and full-color illustrations to encourage evaluation of contrasting perspectives.

The ancient Roman philosopher Lucius Annaeus Seneca wrote, "It is quality rather than quantity that matters." More than just a collection of content, the *Compact Research* series is simply committed to creating, finding, organizing, and presenting the most relevant and appropriate amount of information on a current topic in a user-friendly style that invites, intrigues, and fosters understanding.

Animal Experimentation at a Glance

Why Animals Are Used

Animals provide living systems for scientists to test their theories, medicines, and treatments without putting human lives at risk. Whether animal experimentation can be eliminated, or the number of animals greatly reduced, is a controversial issue.

Medical Progress

Over the years animal testing has been used to develop vaccines against polio, rabies, influenza, and other diseases, as well as medical treatments, surgeries, organ transplants, and new drugs. Some people believe that these breakthroughs could have been made without the use of animals and even say that animal experimentation has hindered scientific progress. Animal research supporters do not agree, insisting that animals are absolutely necessary in order to develop lifesaving drugs, treatments, and medical procedures.

Product Testing

In addition to medical research, animals are used in laboratories to test the toxicity of cosmetics, toiletries, fragrances, and household products before these products are released to the public for human use. While this is not mandated by law, manufacturers of these products are required to perform extensive testing and are encouraged to use animals if necessary.

Types and Quantities of Animals Used

According to the U.S. Department of Agriculture (USDA), 1,177,567 animals were used in research during 2005. This number excludes mice, rats, birds, reptiles, amphibians, and fish, as well as animals owned by federal research agencies and creatures that are kept specifically for breeding. Mice and rats are used in about 90 percent of all research experiments.

Animal Use in Education

The Humane Society of the United States estimates that 6 million vertebrate animals, and about the same number of invertebrates, are dissected in American high schools each year. Unknown quantities of animals are also used for dissection in elementary schools, middle schools, and colleges. Many veterinary schools and some medical schools use live animals as part of their students' training. Whether the use of animals in education is beneficial to students is a controversial issue.

Legislation

Animals used for research in the United States are protected by the Animal Welfare Act (AWA), which sets standards of care that apply to the creatures' housing, feeding and watering, cleanliness, ventilation, and medical needs. The law protects most warm-blooded animals that are bred for commercial sale, used in research, transported commercially, or exhibited to the public, with the exception of those excluded in USDA reporting.

Private Protection of Animals

Animal advocacy groups such as People for the Ethical Treatment of Animals (PETA), Physicians Committee for Responsible Medicine (PCRM), and the Humane Society use their voices and power to pressure researchers to stop experimenting on animals, and to influence politicians to pass laws that offer animals greater protection. Most people in these groups advocate peaceful protest against animal experimentation. Extremist groups such as Animal Liberation Front (ALF) believe that violent tactics are acceptable if necessary to protect animals.

Treatment of Research Animals

Scientists who use animal experimentation say that every possible effort is made to provide humane care for animals kept in laboratories. This is an issue of controversy because rats and mice, which are used for 90 percent of scientific research, are not protected under the AWA. Also, some facilities have been cited for reporting erroneous numbers of creatures used in research, as well as neglecting or abusing animals.

Alternatives

Many scientists, even those who support animal experimentation, are searching for alternative ways to conduct research in their laboratories. This is known as the 3 Rs: replacement, reduction, and refinement.

Overview

"Knowledge gained from animal research has extended and improved the quality of human and animal life in countless ways."

—Oregon Health & Science University, "Why Use Animals in Research?" Oregon National Primate Research Center. http://onprc.ohsu.edu.

"Defenders of animal experimentation have done a poor job of making their case for the benefits of such research, often relying on sweeping generalizations or unsupported claims."

—Martin Stephens, quoted in "Doubts Raised over Value of Animal Research," Humane Society of the United States, April 1, 2004. www.hsus.org.

For as long as medical science has existed, animal experimentation has played a major role in research. Over the years scientists have used animal experiments to aid in the development of numerous drugs and medical procedures, as well as to gain a better understanding of how the human body works. As far back as the 2nd century, a Roman physician named Galen of Pergamum wanted to learn more about the human body by studying anatomy. Because the Church forbade human dissection, he experimented on pigs, dogs, apes, and other animals, with the goal of demonstrating that veins carry blood, not air, as scientists previously believed. He was one of the first scientists to relate his findings in animals to human beings.

In the early 1600s no one fully understood the function of the heart, and most scientists believed that the lungs were responsible for moving blood throughout the body. English physician William Harvey questioned these beliefs and began experimenting with a wide range of animals to study blood circulation. By dissecting dead animals, as well as observing the beating hearts of living animals, Harvey drew conclusions about human physiology. He determined that it was the heart, not the lungs, that caused blood to travel in a circular pattern throughout the body. He later published his findings in a work entitled *On the Movement of the Heart and Blood in Animals*, in which he discussed the value of vivisection, or the act of cutting into animals. Over the years the word *vivisection* took on a broader meaning and is now often used to describe animal experimentation in general.

“For as long as medical science has existed, animal experimentation has played a major role in research.”

Animal Cruelty in the Past

Today, animal experimentation is governed by laws that help protect laboratory animals by ensuring that they are treated humanely and suffer as little as possible—but that was not always the case. Early animal experiments were barbaric and cruel; and because anesthesia did not yet exist, the creatures suffered excruciating pain. In 1824, for example, French physiologist François Magendie was in London as a guest of English chemist and physicist William Hyde Wollaston. During Magendie's time in England, he performed dissection demonstrations on the brains of living dogs, during which, writes author Deborah Rudacille, “The man tortured animals in public, slicing into living flesh as if it were a piece of mutton, as the bound beasts screamed in agony.”[1] People in England were sickened by the brutal procedures, and their outrage led to the formation of the Society for the Prevention of Cruelty to Animals in Great Britain, as well as the first animal protection laws.

How and Why Animals Are Used

Scientists use animals for research and experimentation because they must test their theories, medicines, and treatments without putting

Research animals are chosen for different reasons, based on the species. For example, chimpanzees have physiological systems that are remarkably similar to humans, making them susceptible to many of the same health problems as people. In most cases, researchers only use primates when a vital scientific question cannot be answered by using other types of animals.

human lives at risk. Also, many animals have relatively short life cycles, which allows them to be studied over their entire life span. Research animals are chosen for different reasons, based on the species. For example, monkeys, chimpanzees, and rabbits have physiological systems that are remarkably similar to humans, so the creatures are susceptible to many of the same health problems as people. Mice are widely used because their immune systems are nearly identical to those of humans.

The cardiovascular and respiratory systems of humans closely resemble those of dogs and pigs, while the neurological systems of cats are very much like humans. Guinea pigs have large ears and inner ear structures that are virtually identical to those of people, so they are often used in research on hearing.

By studying creatures' living systems, scientists gain a better understanding of the workings of the human body, as well as learn about the causes and progression of infection, genetic disorders, and human disease. Researchers use animals to broaden their understanding of human conditions such as clinical depression and alcoholism, as well as drug addiction, fetal alcohol syndrome, and allergies.

Even if researchers are not in favor of using animals for experimentation, those who develop prescription medicines, over-the-counter drugs, or vaccines are required by law to do so. The U.S. Food and Drug Administration (FDA) requires that any new drug be tested on living animals to determine its effect on the human body, how it is broken down chemically, whether it has toxic side effects, and how safe it is at different dosages.

Quantity of Animals Used in Research

The U.S. Department of Agriculture (USDA) reports that 1,177,567 animals were used in research during 2005, including 245,786 rabbits, 221,286 guinea pigs, 176,988 hamsters, 66,610 dogs, 58,598 pigs, 22,921 cats, and other various species. These totals do not include mice, rats, birds, reptiles (such as turtles and snakes), amphibians (such as frogs), and fish. Also not counted in the USDA's reporting are animals owned by federal research agencies, animals that are kept specifically for breeding, or those used in agricultural research. Because of such discrepancies, the actual number of live creatures used for experimentation each year is unknown, although the quantity is likely in the tens of millions.

"The actual number of live creatures used for experimentation each year is unknown, although the quantity is likely in the tens of millions."

Animal Research and Human Health

Scientists have used animal research to develop vaccines that protect people against such deadly diseases as polio, smallpox, diphtheria, tetanus, rubella, whooping cough, and influenza. Animal research has also paved the way toward the development of antibiotics, anticoagulants that prevent blood clots, medicines to manage high blood pressure or lower cholesterol, and thousands of other lifesaving drugs. By studying mice and rats, scientists have gained knowledge about breast cancer, including the genetic and environmental factors that affect it. Research using dogs, sheep, and pigs helped scientists make advances in organ transplants, including the development of the antirejection drugs patients must take after transplant surgery. Chemotherapy for cancer treatments was first tested on animals, as were artificial heart valves, organ transplants, and knee and hip replacement surgeries. Pigs have helped scientists develop treatments for burn victims, as pig skin is often used as a temporary bandage that can protect against infection while new human skin is generated.

Product Testing for Safety

Although the FDA does not mandate that cosmetics, toiletries, fragrances, and common household products be tested on animals, the agency does require extensive safety testing and encourages manufacturers of these products to use animals if absolutely necessary. According to the Michigan Society for Medical Research, "Manufacturing must substantiate the safety not only of finished products, but also of every ingredient the products contain. The only way to obtain this information is through testing on animals."[2]

Researchers also experiment with animals to determine the possible consequences if products are misused. They use their findings to develop antidotes for poisonous substances that may be accidentally ingested, inhaled, or splashed in someone's eyes, or when someone overdoses on a particular drug. Poison centers, which receive more than 1 million calls each year about such emergencies, are able to tap into this information when they are contacted. Animal studies with a chemical called ethylene glycol, which is used to make antifreeze and deicing products for automobiles, airplanes, and boats, helped researchers understand the different levels at which the chemical was most toxic. According to the

American Association of Poison Control Centers in Washington, D.C., eye exposures to ethylene glycol account for about 10 percent of calls to U.S. poison centers. Research with the chemical enabled scientists to develop guidelines for medical personnel that help them in the event that someone is exposed to it.

Are Animals Necessary for Product Testing?

Fewer animals are used for testing household products and cosmetics than five years ago, and some companies have discontinued the use of animals altogether. But animal experimentation is still widely accepted as a way of testing product safety. More and more people are speaking out against these tests because of the suffering they inflict on animals, as well as questioning whether the tests are effective. As the National Anti-Vivisection Society states, "Despite the fact that they have been animal-tested, these products are no less deadly if a person eats or drinks them accidentally."[3] Another argument against animal experimentation is that laboratory animals are often subjected to unnecessarily repetitive tests and are given much larger doses of toxic chemicals and other substances than humans would ever be subjected to. In tests to evaluate the cancer-causing properties of the artificial sweetener saccharin, for example, rats were fed a dose equivalent to a human consuming more than a thousand cans of diet soda.

> **"Animal experimentation is still widely accepted as a way of testing product safety."**

Tests such as those used to determine the toxicity of chemicals are known as Lethal Dose Fifty Percent, or LD50. Animals are force-fed, or forced to inhale, poisonous substances to determine at what dose half of them will die. Sometimes the animals die quickly after being poisoned to death by high levels of toxins; in other cases the creatures die slowly as their internal organs are gradually destroyed. Opponents of animal experimentation say that not only are LD50 tests cruel, they are highly unreliable because the results obtained from them are inconsistent from species to species or even animal to animal.

Another common laboratory test, known as the Draize test, is used in two different ways. The Draize skin irritant test gauges the harmfulness of numerous products, from chemicals used in cosmetics to drain cleaners, by observing whether the substances burn the skin of animals or corrode their skin, which means it is destroyed beyond its ability to heal. The second type, the Draize eye irritant test, is used to determine how much damage a caustic substance causes to an animal's eyes. Once a creature's head is confined in a head-holding device, the test substance is poured in its eyes and researchers observe and record the resulting irritation, swelling, hemorrhaging, and damage over several weeks. Because the Draize test is perceived as cruel, as well as obtaining highly variable results, many scientists and laboratories do not endorse it. Stephen R. Kaufman, an ophthalmologist at Bellevue Hospital in New York City, is strongly against Draize eye irritant tests and refuses to use any data from them in treating his patients. Kaufman calls the tests "scientifically unsound," primarily because there are vast differences between rabbit eyes and human eyes. He also explains the risk for humans: "Reliance on this test is in fact dangerous, because the animal data cannot be reliably extrapolated to man. Substances 'proven' safe in lab animals may in fact be dangerous to people."[4]

Animal Use in Education

Animals have been used in science classrooms of secondary schools, colleges, and universities at least since the 1920s. Frogs, fetal pigs, earthworms, and other creatures are usually purchased from biological supply houses, which kill and preserve the animals in chemicals. Students then perform dissections to learn about the creatures' anatomy and physiology. In medical and veterinary education, students dissect dead animals and sometimes use live animals to practice surgical techniques and other procedures before working on humans. Although animal use in education is still prevalent throughout the United States, more and more students have begun objecting to the practice and speaking out about their beliefs. Nine states have passed laws that give students the ability to opt out of dissection and choose to participate in alternative methods (such as computer simulations) instead.

How Animal Experimentation Is Regulated

Most research facilities that use animals are required to be registered with the U.S. Department of Agriculture (USDA) and are protected by the Animal Welfare Act (AWA). Originally passed in 1966, the AWA sets standards of care for laboratory animals that apply to their housing, feeding and watering, cleanliness, ventilation, and medical needs. Facilities that use only mice, rats, birds, or cold-blooded animals, however, are not required to register with the USDA, so these creatures are excluded from AWA protection.

The AWA was amended several times after its passage. Amendments in 1985 mandated that research facilities provide an adequate physical environment for dogs and an environment that promotes the psychological well-being of primates. The amendment also specified that pain and distress for animals must be minimized during experimental procedures and that alternatives to such procedures must be considered whenever possible. Another change in 1985 was the requirement that research organizations establish an Institutional Animal Care and Use Committee (IACUC) of no less than three members, one of whom is a licensed veterinarian. The committee is charged with overseeing the work of the institution and ensuring that animals used in research are treated properly and responsibly. The AWA also sets stringent guidelines for researchers seeking permission to use animals in research and requires the USDA to make random, unannounced visits to research facilities to ensure that they are in compliance with federal regulations.

“Facilities that use only mice, rats, birds, or cold-blooded animals . . . are not required to register with the USDA, so these creatures are excluded from AWA protection.”

In addition to being governed by the AWA, scientists who seek federal funding for their research must also adhere to standards for the humane care of animals set forth by the National Institutes of Health (NIH). To ensure compliance with its requirements, the NIH makes periodic site visits to research institutions and laboratories and may revoke or withhold federal grants or contracts awarded for research if facilities are in violation of its policies.

Is Animal Experimentation Humane?

Although many scientific laboratories go to great lengths to ensure that animals are well cared for and treated humanely, in some cases animals have been neglected or suffered great pain without any drugs to relieve it. Also, a number of laboratories have been found in violation of the AWA, either through erroneous reporting or disregard for the laws that govern animal welfare. Another concern of animal experimentation opponents is that rats and mice—which are used for at least 90 percent of all research experiments—are not protected by the AWA. As a result, it is unknown how these creatures are treated or how much pain and/or distress they may suffer.

When Animal Research Has Failed

While it is true that animal experimentation has contributed toward significant breakthroughs in medical science, some humans have been harmed because of flawed findings. One such incident involved Thalidomide, a drug that was commonly prescribed for pregnant women suffering from morning sickness. Before Thalidomide was released to the public, it had been tested on rats, mice, rabbits, dogs, hamsters, primates, guinea pigs, and cats; and researchers determined that it was totally safe for humans. But in the late 1950s and early 1960s an estimated 10,000 children were born with birth defects that were directly connected to the drug.

> **“One of the most pivotal cases involving the gross abuse of research animals is known as the Silver Spring Monkey case.”**

Another case involved Vioxx, a drug that was developed for the treatment of arthritis. When tested on animals, Vioxx appeared to be safe and potentially beneficial in treating heart disease as well as arthritis. After it was released for human use, however, the drug caused more than 100,000 heart attacks and strokes among people in the United States. One FDA representative was quoted as saying that the Vioxx incident was the "single greatest drug-safety catastrophe in the history of the world."[5]

Animal testing also steered scientists in the wrong direction with regard to hormone replacement therapy (HRT), or synthetic hormones developed for women who have depleted estrogen levels due to menopause. When the drugs were tested on monkeys, researchers determined that HRT lowered the creatures' risk of heart attack and stroke. But when prescribed for millions of women, HRT was shown to have the opposite effect—actually increasing their risk of heart disease, breast cancer, and stroke. An August 2003 issue of the British publication *Lancet* reported that in a 10-year period, an estimated 20,000 cases of breast cancer in Great Britain were the direct result of HRT, along with thousands of heart attacks and strokes.

The Silver Spring Monkeys

One of the most pivotal cases involving the gross abuse of research animals is known as the Silver Spring Monkey case, which occurred in 1981 and helped form animal rights groups and amendments to the AWA. Alex Pacheco, who at the time was a student at George Washington University, interviewed for a summer position at the Institute for Behavioral Research (IBR) in Silver Spring, Maryland. During a meeting with psychologist Edward Taub, the IBR director, Pacheco learned that Taub's NIH-funded research involved surgically crippling monkeys by cutting their sensory nerves. Taub explained that his objective was to see how well the monkeys functioned or reacted to electric shock and other such stimuli after they had been paralyzed. Pacheco then toured the facility with Taub and was stunned at the filth and abuse that he saw around him. He described animal feces piled in the bottom of cages, urine and rust encrusted on the bars of cages and other metal surfaces, and distraught monkeys confined in tiny, cramped cages. Many of the creatures were so confused at the numbness in their limbs that they had chewed and mutilated them. The creatures were also starving and desperate to forage for food, as Pacheco explains:

> In their desperation to assuage their hunger, they were picking forlornly at scraps and fragments of broken biscuits that had fallen through the wire into the sodden accumulations in the waste collection trays below. The cages had clearly not been cleaned properly for months. There were no dishes to keep the food away from the faeces, nothing for the animals to sit on but the jagged wires of the old cages, nothing for them to see but the filthy, faeces-splattered walls of that windowless room. . . . It was astounding that Taub and the other researchers expected to gain any reproducible, let alone reliable, data from these animals, considering the condition of the animals themselves and of the colony and surgery rooms.[6]

In order to closely observe what was happening at the laboratory, Pacheco accepted a volunteer position and began working at night and on weekends to record his findings and take photographs. By September 1981 he believed he had enough evidence for Taub to be charged with animal cruelty, and he notified the Silver Spring police, who conducted the first raid in the United States on a research facility. The monkeys were seized by police officers, and Taub was arrested for animal cruelty. His was tried and convicted, but his conviction was later overturned when an appellate court ruled that his federal funding exempted him from the requirements of Maryland's animal cruelty laws. Positive outcomes also resulted from the Silver Spring Monkey case, however. The massive publicity it generated heightened people's awareness of the treatment of animals used in research, and it led to the founding of People for the Ethical Treatment of Animals (PETA) by Pacheco and Ingrid Newkirk. Also, the AWA was amended in 1985 to strengthen protection for primates and dogs.

Speaking Out for Animals

Animal rights groups such as PETA, the Humane Society, Physicians Committee for Responsible Medicine (PCRM), and the Anti-Vivisection Society are known for educating the public about how animals are used in research and how they suffer, as well as putting pressure on lawmakers and research laboratories to slow or halt animal experimentation. Most people who belong to such groups advocate peaceful protest rather than lawbreaking or violence, because their objective is greater protection for animals. There are some groups, however, that believe the only way to stop what they perceive as egregious cruelty to animals is to use extreme measures. The Animal Liberation Front (ALF), which is known as one of the most radical animal

> **"In spite of the progress that continues to be made with alternative research, many scientists believe it is impossible to know when, or even if, animals can ever be completely replaced with other methods."**

Most people who belong to animal rights groups advocate peaceful protest. Here a woman wears a monkey costume in a cage protesting a pharmaceutical company's use of animals for drug research. There are some groups, however, that believe the only way to stop what they perceive as egregious cruelty to animals is to use extreme measures.

rights groups, has criticized PETA and other organizations for being too "soft" on researchers who use animals in their experiments. Members of these extremist groups have stalked and threatened researchers and their families, made harassing telephone calls, and attacked people in their homes. They have also issued death threats, bombed research laboratories, and set fires. This sort of violence led to the passage of the Animal Enterprise Terrorism Act in November 2006, which is intended to help federal authorities better investigate and prosecute individuals who commit violent crimes in the name of animal liberation.

Alternatives to Animal Experimentation

In an effort to address the growing concerns over animal experimentation, more and more scientists are looking for alternative ways to conduct research in their laboratories. Scientists are constantly looking into technology to replace or reduce animals used in research, as well as finding new ways to use human cells or tissue for experiments, rather than animals. This is known as the three Rs: replacement, reduction, and refinement, and it is widely supported even by researchers who use animals.

Yet in spite of the progress that continues to be made with alternative research, many scientists believe it is impossible to know when, or even if, animals can ever be completely replaced with other methods. The National Association for Biomedical Research explains: "No responsible researcher believes that the technology exists today, or in the foreseeable future, to replace the use of animals altogether in biomedical research."[7] Still, scientists throughout the world are committed to their ongoing search for ways to use animals only if necessary; to eliminate the creatures' pain and suffering as much as possible; and to reduce the number of animals used.

Do the Benefits of Animal Experimentation Justify Its Use?

"Americans must decide whether they support animal research or animal rights. . . . We can't let a potential treatment for AIDS fall victim to their specious rhetoric."

—Joseph E. Murray, "Laboratory Animal Farm," 2001.

"Usually, [animal experimenters] simply point to the supposed human benefits and argue that the ends justify the means, though they rarely substantiate their claims with scientific evidence."

—Christopher Anderegg et al., "A Critical Look at Animal Experimentation," 2006.

Whether animals should be used for research is one of the most controversial topics known to science. Many people, including scientists and medical professionals, are cautious about the animal experimentation issue. They believe that the idea of living creatures suffering and being sacrificed for the sake of humanity is unfortunate but understandable, in that animal research is a necessary part of medical and scientific progress. That is the perspective of Ron Otten, a microbiologist who says, "There has always been some inner conflict for me, but a lot of that was put to rest during my travel in east Africa. I have seen the valuable role that animal research can fulfill—not only in the field of HIV/AIDS research but across the spectrum. The bottom line is, we need this. . . . It has its place."[8]

Progress and Problems

No one questions that animal experiments have played a major role in the development of lifesaving vaccines, medicines, surgeries, and revolutionary treatments. For example, noted French chemist Louis Pasteur made medical history in 1885 by creating a vaccine for rabies. Pasteur grew the rabies virus in live rabbits and then killed the animals and removed their diseased spinal cords. After drying and crushing the spinal cords, he mixed the powdered substance with a chemical solution and then injected it into dogs. He had not planned to test the vaccine on humans—but when a nine-year-old boy was bitten by a rabid dog, Pasteur knew he would die without immediate treatment. He injected the boy with multiple doses of his experimental vaccine and, in the process, saved his life.

> **“No one questions that animal experiments have played a major role in the development of lifesaving vaccines, medicines, surgeries, and revolutionary treatments.”**

In the early 1950s animal testing enabled scientific researcher Jonas Salk to develop the polio vaccine. Before the vaccine existed, poliomyelitis was a terrifying epidemic that killed or paralyzed thousands of people, many of them young children. Salk created an experimental vaccine and then injected it into live monkeys. Within three weeks, blood samples taken from the creatures showed that they were developing antibodies, or natural disease-fighting agents created by their immune systems. Salk tested his vaccine on a small group of humans, and they, too, developed antibodies. The vaccine was made available to the public, and incidents of polio began to drop dramatically. In the late 1950s Albert Sabin developed another polio vaccine that could be taken orally, and he also used laboratory animals. Sabin later wrote about the decades of research leading up to the creation of his vaccine, saying he had performed experiments on “many thousands of monkeys and hundreds of chimpanzees.”[9] From 1950 to 1980 incidents of polio in the United States dropped from 33,300 to 9, and the disease was completely wiped out in the United States by the end of the twentieth century.

Animal experimentation opponents do not deny that medical progress has been made throughout the years. Their question is whether the same results might have been achieved without using animals. Did the progress occur because of animal experimentation or in spite of it? Some people are convinced that medical progress has been hindered by animal experimentation because of flawed results, such as with cancer research. Of 20 substances that were shown not to cause cancer in humans, 19 did cause cancer when tested on mice. Cancer treatments that showed great promise in treating mice were not effective at all when tested on people. As a result of these and other failed experiments, some scientists believe that much of the cancer research performed on animals has been futile, as Richard Klausner, former director of the National Cancer Institute, says: "The history of cancer research has been a history of curing cancer in the mouse. We have cured mice of cancer for decades, and it simply didn't work in humans."[10]

Is Animal Experimentation Moral?

The strongest opponents of animal experimentation believe that it is inherently wrong, and no matter what results it has achieved, it should be stopped because it is immoral. They insist that humans and animals have an equally innate right to live, and therefore people do not have the right to sacrifice any living creature's life for the sake of prolonging or enhancing human lives. As Chris DeRose, founder of the group Last Chance for Animals writes, "If the death of one rat cured all diseases, it wouldn't make any difference to me."[11]

> **"Of 20 substances that were shown not to cause cancer in humans, 19 did cause cancer when tested on mice."**

The debate over whether humans are the superior species has been raging for centuries. When arguing in favor of human superiority, many people base their beliefs on religious views, quoting verses from the Bible such as Genesis 1:28: "And God blessed them and told them, 'Multiply and fill the earth and subdue it; you are the masters of the fish and birds and all the animals."[12]

In the 1850s British scientist Charles Darwin challenged accepted religious beliefs with his book called *The Origin of Species by Means of*

Natural Selection, in which he presented the theory of evolution. After more than 20 years of study, Darwin refuted the belief that humans were at the top of an evolutionary progression of animal species; rather, he wrote that humans are a highly functioning species of animal that has survived, adapted, and changed over thousands of years—in other words, humans were not superior to animals, they were just a higher species of animal. Yet even though Darwin drew closer parallels between humans and animals than most any scientist had done before, he was a strong supporter of animal research, as he explained in a letter to a Swedish professor in 1881: "I know that physiology cannot possibly progress except by means of experiments on living animals, and I feel the deepest conviction that he who retards the progress of physiology commits a crime against mankind."[13] Despite Darwin's stance in favor of animal experimentation, his writings were influential in the development of anticruelty laws as well as the formation of animal rights groups.

> **"Animal suffering is a huge concern not only for animal rights advocates but also many scientists and researchers—even those who use animals in experiments."**

Scientist and professor Stuart Derbyshire insists that humans are the superior species, and therefore, animal research is valuable and necessary. He talked about this in a speech given at the 2002 Edinburgh Book Festival: "Animal welfare is not the aim of animal experiments, human welfare is. To defend animal experimentation it is necessary to champion humanity. . . . For me, humans trump animals every time and it is right and proper that animals be sacrificed for the cause of humanity. . . . Humans and animals are not on the same scale."[14]

How Much Do Animals Suffer?

Opponents of animal experimentation often cite the creatures' suffering as the main reason they believe such practices are wrong. They believe that humans have no right to inflict pain on animals in the name of science. In fact, animal suffering is a huge concern not only for animal rights advocates but also many scientists and researchers—even those who use animals in experiments. In its *Animals Used in Research* report, the USDA provides information about the pain and/or distress suffered by animals in research facilities. The report states that during 2005 more than half of the animals used did not suffer at all. About 36 percent suffered pain and distress but were given anesthesia or other drugs to help relieve their pain. About 7 percent of the animals suffered pain or distress and were not given drugs for relief, either because the experiments involved the study of pain, or because anesthesia would interfere with the test results. Since rats, mice, birds, and some other animals are not protected by the AWA or included in the reporting process, no data is available about the pain and/or distress these creatures might suffer during research and testing.

Claire Andre and Manuel Velasquez of the Markkula Center for Applied Ethics at Santa Clara University address the issue of animal suffering in a paper entitled "Of Cures and Creatures Great and Small" in which they say the morally relevant question about animals is, can they suffer? "And, animals do in fact suffer," they write, "and do in fact feel pain."

> The researcher who forces rats to choose between electric shocks and starvation to see if they develop ulcers does so because he or she knows that rats have nervous systems much like humans and feel the pain of shocks in a similar way. Pain is an intrinsic evil whether it is experienced by a child, an adult, or an animal. If it is wrong to inflict pain on a human being, it is just as wrong to inflict pain on an animal.[15]

Primates in Animal Research

Primates have physiological systems that are much like humans, so they are used in many different types of experiments and tests. In most cases, researchers only use primates when it is absolutely necessary, such as when a vital scientific question cannot be answered by using other types of animals. Chimpanzees, which are considered a higher primate species than monkeys, are used primarily for research on hepatitis and HIV, the virus that causes acquired immunodeficiency syndrome (AIDS). Monkeys are also used for HIV experimentation, as well as drug testing and research into malaria and other infectious disease. According to the USDA, a total of 57,531 primates were used for research purposes in 2005.

One organization that specializes in primate research is the Yerkes Regional Primate Research Center in Atlanta, Georgia. In a medical breakthrough, Yerkes researchers studying AIDS developed a vaccine that prevented monkeys from contracting HIV. They are hopeful that their discovery will lead to the creation of a vaccine that can protect humans and eventually eliminate the deadly disease. Through experiments with rhesus monkeys, scientists at the Oregon National Primate Research Center discovered a monkey version of the herpes virus, which is helping them learn more about how the human virus causes a type of cancer called Kaposi's sarcoma in AIDS patients. At an animal testing laboratory in Oxford, England, scientists used about 150 monkeys in experiments to develop deep-brain-stimulation treatments that help control body tremors in patients who have Parkinson's disease. According to Mark Walport, director of the organization that sponsors the research, these types of experiments have played a vital role in medical science. "No one likes doing primate experiments," he says, "but some research can only be done on monkeys."[16]

> **"At one time chimps inhabited 25 African countries; today, however, they are either nearly or completely extinct in 13 of those countries."**

Of all the animals used in research, primates are the source of the greatest controversy. Opponents of using these creatures for research and testing point out that they are more similar to humans than any other creature on Earth, and thus, it is inherently wrong for them to be sacrificed. Research has shown that primates possess many characteristics that were once thought to be uniquely human, such as feeling fear, anxiety, boredom, frustration, and stress. Famed researcher and author Jane Goodall has studied primates for many years, and she says that to subject the creatures to physical or psychological suffering is not only unnecessary but also unethical. Like many other people who are against the use of primates in research, Goodall points out that they have complex brains and nervous systems, complex minds, live in complex societies, and show emotions that in some cases are very similar to those of humans. "In view of these physiological and anatomical similarities," she writes, "it is sad to find that the equally striking similarities between ourselves and these apes in the sphere of behaviour, emotional expression and intellectual performance have been largely disregarded or even denied by many of the researchers who use the living bodies of chimpanzees in their laboratories."[17] One major concern of those who are opposed to the use of chimpanzees in research is that they are an endangered species. At one time chimps inhabited 25 African countries; today, however, they are either nearly or completely extinct in 13 of those countries.

An Ongoing Battle

People have myriad opinions about the animal experimentation issue, especially about whether the benefits to humans are worth sacrificing the lives of millions of animals. Those who are against the practice argue that humans are not superior to animals and have no right to inflict pain and suffering on living, breathing, feeling creatures. Those who support it cite years of medical progress and improved health for people all over the world.

Primary Source Quotes*

Do the Benefits of Animal Experimentation Justify Its Use?

“Some people may be repulsed by the very idea of vivisection, but the fact remains that animal research is an essential component of medical research.”

—*Deseret Morning News*, “Intimidation Is Ineffective,” editorial, October 27, 2006. http://deseretnews.com.

The *Deseret Morning News* is a Salt Lake City, Utah, daily newspaper.

“For now, science marches on with little government oversight, little or no public input, and seemingly little consideration for ethical concerns.”

—Tamiko Thomas, “The Chicken Called; She Wants Her Legs Back,” *AV Magazine*, Summer 2005. www.aavs.org.

Thomas is an animal scientist who works for the Humane Society of the United States.

Bracketed quotes indicate conflicting positions.

* Editor's Note: While the definition of a primary source can be narrowly or broadly defined, for the purposes of Compact Research, a primary source consists of: 1) results of original research presented by an organization or researcher; 2) eyewitness accounts of events, personal experience, or work experience; 3) first-person editorials offering pundits' opinions; 4) government officials presenting political plans and/or policies; 5) representatives of organizations presenting testimony or policy.

“Millions of humans would suffer and die unnecessarily if animal testing were prohibited. But this is exactly what PETA and other ‘animal rights’ organizations seek.”

—Alex Epstein, “Animal Rights Movement’s Cruelty to Humans,” *Capitalism Magazine*, August 15, 2005.

Epstein is a writer for the Ayn Rand Institute in Irvine, California.

“We do not believe that the utilitarian philosophy of ‘sacrificing’ a few for the benefit of many is morally defensible regardless of the species in question.”

—Association of Veterinarians for Animal Rights (AVAR). www.avar.org.

AVAR’s mission is to pursue rights for all nonhuman animals by educating the public and veterinary professionals about various issues, including the use of animals in research and testing.

“Although nobody likes the idea of any research using non-human primates, if it comes to a choice between regulated studies on a few animals and a treatment for an incurable disease affecting hundreds of thousands of people, most people reluctantly make the same choice.”

—Chris Higgins, quoted in Rebecca Morelle, “UK Experts Back Primate Research,” BBC News, December 12, 2006. http://news.bbc.co.uk.

Higgins is the director of the Medical Research Council Clinical Sciences Centre in London.

"Even if it should be proved that human beings benefit directly from the suffering of animals, its infliction would nevertheless be unethical and wrong."

—Hugh Dowding, quoted in BBC News, "Viewpoint: Animal Rights," September 7, 2004. http://news.bbc.co.uk.

Dowding, who died in 1970, was an officer in the Royal Air Force and Great Britain's air chief marshal.

"When the medical community tells you that we use chimps because they share 99 percent of our genes, I say that is a fallacious argument. We share 58 percent of our genes with bananas and no one is testing heart disease on bananas."

—C. Ray Greek, quoted in Ahmed Hamid, "Author Claims 'U' Experiments on Animals Ineffective," *Michigan Daily*, January 29, 2001.

Greek is an anesthesiologist who is president of Americans for Medical Advancement, an organization that is opposed to animal research.

"Depriving sick human beings of the benefits of animal research is inhumane and reprehensible."

—American Physiological Society, *Statements on Animal Usage*, October 1987. www.the-aps.org.

The American Physiological Society, an organization that supports animal research, is devoted to fostering education, promoting research, and sharing scientific information.

"A rat is a pig is a dog is a boy."

—Ingrid Newkirk, quoted in Steven Milloy, "Laboratory Animal Farm," Fox News.com, April 4, 2001. www.foxnews.com.

Newkirk is the cofounder of People for the Ethical Treatment of Animals (PETA).

"Scientific research requiring laboratory animals continues to result in spectacular achievements that have advanced our understanding of life and treatment of disease."

—American College of Laboratory Animal Medicine, *Position Statement on Animal Experimentation.* www.aclam.org.

The American College of Laboratory Animal Medicine is an organization that is dedicated to advancing the human care and responsible use of laboratory animals.

"The reliance on animal-modeled research, as well as other pseudoscientific endeavors, harms rather than helps humans, and prolongs human suffering by inhibiting medical progress."

—Americans for Medical Advancement. www.curedisease.com.

Americans for Medical Advancement is a scientific organization that opposes the use of animals as a means of seeking cures and treatments for human disease.

Do the Benefits of Animal Experimentation Justify Its Use?

- According to the U.S. Department of Agriculture (USDA), a total of **1,177,586 animals** were used for research purposes in 2005, **57,531** of which were primates. This is an increase over 2004, when **54,998** primates were used in research.

- Great Britain, Austria, the Netherlands, New Zealand, and Sweden have all banned the **use of great apes** in research, including chimpanzees, gorillas, and orangutans.

- The majority of **animals used in laboratories** have been bred specifically for research.

- Research with animals has led to vaccines that protect people against **polio, diphtheria, measles, whooping cough, tetanus, rabies, influenza,** and numerous other diseases.

- In a poll conducted for the Humane Society of the United States, **75 percent** of people surveyed disapprove of experiments that subject animals to severe pain and distress.

- In a 2004 poll commissioned for the American Anti-Vivisection Society, **67 percent** people surveyed said they considered it unethical to issue patents on animals. In the same survey, 85 percent of the respondents were not aware that animals could be patented.

Animal Experiments Have Led to Nobel Prize–Winning Accomplishments

This chart shows some of the Nobel Prize–winning scientists who have been recognized for major work in medicine that resulted from their research with animals.

Year	Scientist(s)	Animal(s)	Contributions made
2006	Andrew Fire, Craig Mello	Worm	Discovery of a fundamental mechanism for controlling the flow of genetic information
2003	Paul Lauterbur, Sir Peter Mansfield	Clam, rat	Imaging of human internal organs with exact and non-invasive methods (MRI)
1995	Edward Lewis, Christiane Nusslein-Volhard, Eric Wieschaus	Fruit fly	Genetic control of early structural development
1990	Joseph Murray, E. Donnall Thomas	Dog	Organ transplantation techniques
1979	Allan Cormack, Godfrey Hounsfield	Pig	Development of computer assisted tomography (CAT scan)
1966	Peyton Rous, Charles Huggins	Rabbit, rat, hen	Tumor-inducing viruses and hormonal treatment of cancer
1954	John Enders, Thomas Weller, Frederick Robbins	Monkey, mouse	Culture of poliovirus that led to development of vaccine
1951	Max Theiler	Monkey, mouse	Development of yellow fever vaccine
1945	Sir Alexander Fleming, Ernst Chain, Sir Howard Florey	Mouse	Curative effect of penicillin in bacterial infections
1934	George Whipple, George Minot, William Murphy	Dog	Liver therapy for anemia
1923	Frederick Banting, John Macleod	Rabbit, dog, fish	Discovery of insulin and mechanism of diabetes
1904	Ivan Pavlov	Dog	Animal responses to various stimuli

Source: Foundation for Biomedical Research, "Nobel Prizes: The Payoff from Animal Research," www.fbresearch.org.

Animals Used in Research in 2005

Animals are used by scientists to test new vaccinations, medicines, theories, and treatments, as well as for testing household products, toiletries, and cosmetics. According to the United States Department of Agriculture (USDA), a total of 1,177,566 animals were used in laboratories during 2005.

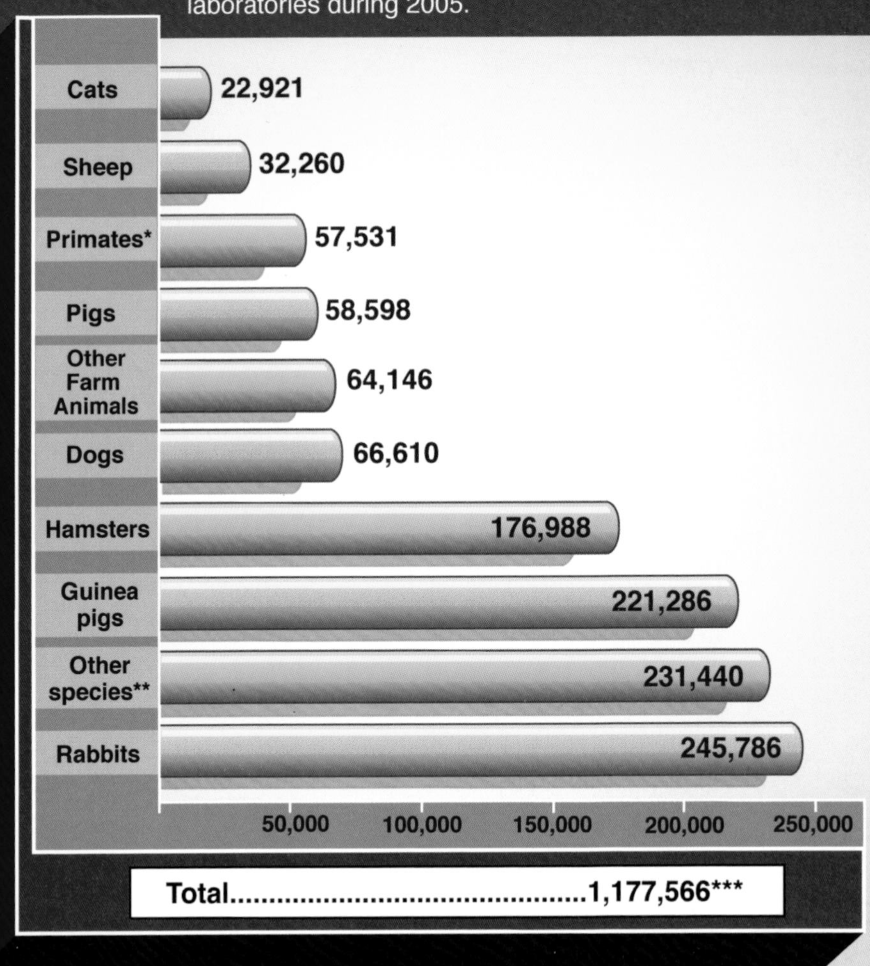

* Primates are a group of animals that includes monkeys and chimpanzees.

**The USDA includes wild animals in this category, such as squirrels, ferrets, bats, and others.

***Mice, rats, birds, and fish are not included in this total.

Source: United States Department of Agriculture, Animal, and Plant Health Inspection Service, "FY 2005 AWA Inspections," www.aphis.usda.gov.

Animal Use in Laboratories

The greatest number of laboratory animals are used for biomedical and exploratory research, followed by drug development, product safety testing, vaccine production, and education. The following illustration shows the nature of animal use in laboratories and a breakdown of the purposes for which animals are used.

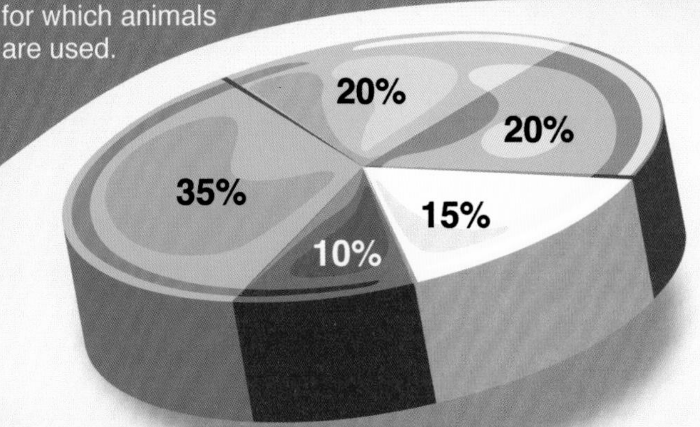

Basic and Applied Research (35%)

- Biomedical research, such as research into cancer and heart disease, infectious disease, and disorders of the brain and nervous system
- Basic exploratory research, namely, experiments that may have no immediate application to human problems but that add to the body of scientific knowledge
- Military experiments
- Agricultural experiments
- Space research

Drug Discovery and Development (20%)

Product Safety and Efficacy (20%)

- Industrial and chemical tests
- Tests of cosmetics, food additives, and pharmaceuticals

Production of Vaccines and Antibodies (15%)

Education (10%)

- Medical and veterinary training
- Classroom dissections and live animal experiments
- School science fairs

Source: National Association for Humane and Environmental Education (NAHEE), youth education affiliate of The Humane Society of the United States, "Science and Conscience: The Animal Experiment Controversy," 2004. www.humaneteen.org.

The Impact of Vaccinations on Reducing or Eliminating Disease

Immunizations are largely a product of the twentieth century and have saved billions of lives throughout the world. Most scientists believe that animal experimentation played a major role in the development of vaccines. The following chart illustrates the impact vaccinations have had in the United States in reducing or eliminating cases and deaths from some of the most deadly diseases.

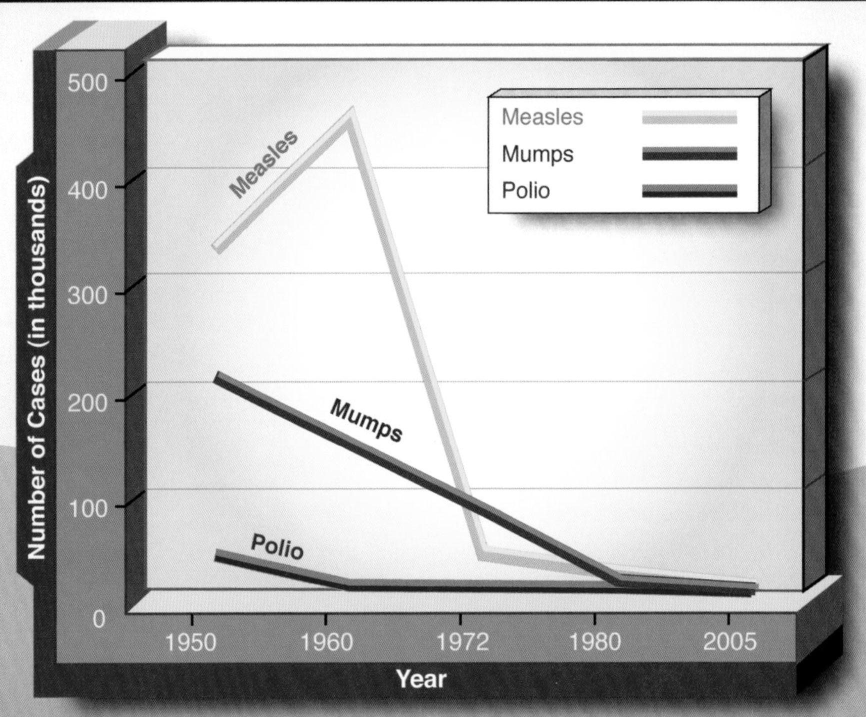

Source: U.S. Department of Health & Human Services and the Centers for Disease Control and Prevention, "Reported Cases and Deaths from Vaccine Preventable Diseases, United States, 1950–2005," December 13, 2006. www.cdc.org, www.immunizationinfo.org.

- In an August 2004 survey of physicians, **82 percent** were concerned that animal research data could be misleading when applied to humans.

- Between 1995 and 2002, AIDS-related deaths in the United States fell by **70 percent** due, at least in part, to new treatments developed through animal experiments.

Should Animal Experimentation Be Used in Education?

“No model, no video, no diagram and no movie can duplicate the fascination, the sense of discovery, wonder and even awe that students feel when they find real structures in their own specimens.”

—Susan Offner, “Dissection Fact Sheet.” www.carolina.com.

“Research consistently shows that cruelty-free alternatives teach concepts of anatomy and biology just as well or better than traditional dissection.”

—Physicians Committee for Responsible Medicine, *Dissection Alternatives.* www.dissectionalternatives.org.

For decades, animals have been used in science classrooms of secondary schools and colleges. Today, animals are used in education in a number of different ways. Creatures such as rabbits, frogs, turtles, hamsters, guinea pigs, birds, and fish are often kept in classrooms so students can learn to care for them as well as observe their behavior. For biology classes where dissection is performed, schools purchase dead animals from biological supply houses, which kill the creatures, preserve their bodies in chemicals such as formaldehyde, and then package them for sale and shipment.

While no official total exists of how many creatures are used in education, the Humane Society of the United States estimates that 6 million vertebrate animals (those having a backbone or spinal column), and about the same number of invertebrates, are dissected in American high schools each year. Unknown quantities of animals are also used for dissection in colleges, middle schools, and elementary schools. The most common creatures dissected are frogs, fetal pigs, and earthworms, but students also dissect mice, rats, turtles, birds, cats, rabbits, fish, salamanders, and bats, among others.

The most common creatures dissected are frogs, fetal pigs, and earthworms, but students also dissect mice, rats, turtles, birds, cats, rabbits, fish, salamanders, and bats, among others.

Changes in Classroom Animal Use

Originally, most animal use in education was limited to the dissection of dead creatures. Then in the 1960s a group of research scientists and educators introduced a new method of study known as the Biological Sciences Curriculum Study (BSCS), which was intended to put greater emphasis on hands-on study. The Humane Society's Jonathan Balcombe explains: "The positive impact of BSCS was that it encouraged students to actually conduct exercises in scientific inquiry and to think more about scientific and biological concepts. The problem was that it asked students to study life by first destroying it.[18] One practice that was a major component of BSCS was pithing, which involved inserting a needle or other sharp object into the skull of a live creature such as a frog or turtle, and then wiggling the object vigorously, thus destroying the animal's brain. Once it was considered brain-dead, students studied its beating heart and lung action and used the creature for physiology experiments.

As these practices became more prevalent, many people became concerned about educational techniques they perceived as destructive. Increased public pressure helped lead to the development of a Code of Practice for secondary school biology education, which was adopted in 1981 by the National Science Teachers Association (NSTA) and the National Association of Biology Teachers. An excerpt of the code reads: "No experimental procedure shall be attempted in mammals, birds, reptiles, amphibians, or fish that shall cause the animal pain or discomfort or that interferes with its health. As a rule of thumb, a student shall only undertake those procedures on vertebrate animals that would be done on humans without pain or hazard to health."[19]

Today, societal concerns have motivated many schools to modify the use of animals in teaching programs, and others are considering such changes. Although many schools still experiment with animals, increasing numbers of schools have discontinued the practice. Len Niebo, who teaches science at a high school in Bergen County, New Jersey, says that the use of dissection in his district has dropped dramatically because of concerns about animal welfare. He explains: "We used to do worms, crayfish, fish, fetal pigs, frogs and even live frogs—destroying the central nervous system so it wouldn't feel pain, and cutting it open to examine the beating heart. No one today would argue that wasn't cruel."[20]

Dissection of living animals does still take place, however. In 2005 a high school substitute biology teacher in Gunnison, Utah, made arrangements for his students to be present during the dissection of a live dog. He believed it would be a good experience for them because they could see the creature's organs actually working. First the dog was sedated with a gas anesthetic, and as students watched, a veterinarian made an incision in its abdomen and removed its digestive system. Following the procedure, the dog was euthanized.

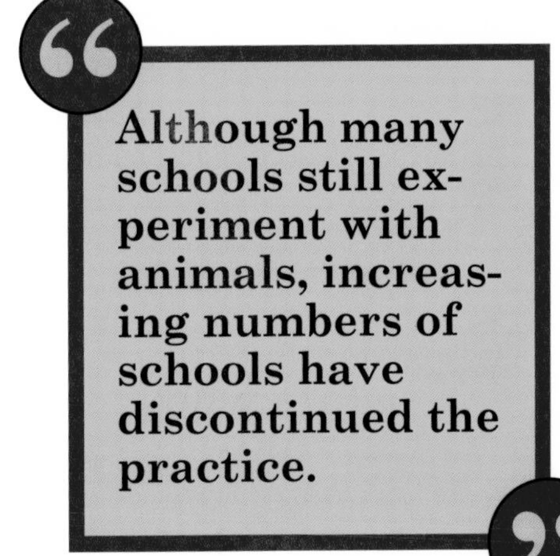

The Purpose of Animals in Education

Scientists and educators who support animal dissection say it provides students with invaluable opportunities to learn about the anatomy and physiology of living things. Students gain an understanding of the animals' internal structures, as well as how their tissues and organs are interrelated. Through hands-on learning, students can become familiar with the complexity of various organisms. Becky Baumgartner, a graduate of Kenston High School in Chagrin Falls, Ohio, explains how valuable this was for her when she was studying biology:

> People say that dissecting animals only makes the dissectors cold, unsympathetic, and unappreciative of life,

> when, actually, just the opposite is true. How can you have great respect for a frog just by looking at a diagram of it? You have to look at the creature up close to really appreciate how it functions. . . . It is mind blowing to see how complex their body systems are—to see how a frog lives, breathes, eats. . . . When it comes to viewing life, you really cannot get any closer than dissection; there is just no substitute for the real thing."[21]

Schools use dissection to teach students about animals' biological systems. At secondary schools in Alpine, New Jersey, seventh graders dissect earthworms, crayfish, fish, frogs, and chickens to learn about the creatures' anatomy and physiology. Fifth graders in Teaneck, New Jersey, dissect pig hearts as part of studying circulatory systems. In some advanced biology classes students dissect pregnant rats to study their reproductive systems. Patricia Lord, a science adviser for Teaneck's schools, explains the importance of dissecting real animals in the biology classroom: "It allows students to feel the delicateness of the tissue and to appreciate the complexity of a living thing. They can inflate the animal's lungs using a pipette to see how lungs work. They can unravel the intestine to see how long it is and how it's packed in. There's so much to learn that way, and they're always amazed."[22]

> **"Scientists and educators who support animal dissection say it provides students with invaluable opportunities to learn about the anatomy and physiology of living things."**

Graduate students studying to become veterinarians also participate in dissection. To learn about anatomy, they dissect cadavers of horses, cows, and dogs. They also use live animals purchased at auctions or slaughterhouses to practice skills such as anesthesia, neutering and spaying, surgery techniques, and euthanasia. The College of Veterinary Medicine at Iowa State University obtains dogs that are no longer needed for breeding by research facilities and are scheduled to be euthanized. Students evaluate the health of the dogs for two to three days and care for them during that time. Then the dogs are anesthetized, and students perform one or more surgical procedures on them. Following the experiments, the creatures are euthanized with an intravenous overdose of barbiturates.

Although living animals are still used at some medical schools, most of them have abandoned the practice. Nearly 90 percent of medical schools no longer use live animals in their programs, and only 13 of the schools still have live animal laboratories. One school that does have such a laboratory is Milwaukee's Medical College of Wisconsin. According to medical student Jeff Tomasini, one physiology class for first-year students involved anesthetizing 60 dogs obtained from local pounds, opening up their chest cavities, examining their beating hearts, and then euthanizing them when the experiments were complete.

Do Educational Animal Experiments Teach Cruelty?

Opponents of animal use in education believe that dissection and other common practices are archaic as well as unnecessary. According to Jane Goodall, requiring students to harm living animals or dissect the bodies of dead animals can undermine their respect for life. She explains:

> Of course it is wonderful to learn about the amazing complexities of even simple creatures, to learn more about the way they work. But there are many other ways in which children can obtain this knowledge . . . ways that will not force sensitive and unwilling students to do something they instinctively hate; ways that do not require the repeated use and discarding of dead bodies of creatures killed in order to teach certain aspects of the life sciences. It would seem in such an instance that the teaching is more about death!"[23]

A three-week summer course at Ohio State University called Spinal Cord Injury Research Techniques was nicknamed "Cruelty 101" by the PCRM. The course is designed for scientists who are involved in spinal cord research, and its purpose is to help them expand their knowledge about such injuries in humans. An estimated 250 rats and mice are used each summer during the course. After performing anesthesia, students cut into the creatures' backs to expose their spinal cords by peeling back skin, muscle, and other tissue. A machine drops a weight onto the animals to crush or tear their spinal cords, and then they are put through a series of behavioral exercises that test the severity of their injuries. The PCRM explains: "The animals' ability to reflexively withdraw limbs and right themselves if dropped upside down, as well as their ability to sense touch and hot temperatures are tested. Their ability to swim, stand, and move on various surfaces . . . are also tested."[24] Aysha Akhtar, who is a neurologist, says that animals used in the class suffer such excruciating pain that some of them chew through their own skin and muscle. She also insists that the class is an ineffective way to learn about spinal cords because human anatomy is so different from that of mice, rats, and other animals.

Conscientious Objectors

Animal experiments are required for science classes in many levels of education, but more and more students are speaking out against such practices and, in some cases, refusing to participate in dissection altogether. There is no federal law that specifically gives them that right, but they are protected under the First Amendment, which reads: "Congress shall make no law respecting the establishment of religion or prohibiting the free exercise thereof." The free exercise clause protects people's rights to believe in, and practice, the religion of their choice. As a result of this clause, if a student personally believes that it is wrong to harm animals, he or she has a constitutional right to refrain from participating in animal experimentation.

This sort of outright dissention is relatively new. In the past, students were threatened with failing grades or expulsion from school unless they participated in all class experiments, including dissection. In 1987 a California student named Jenifer Graham publicly challenged the rules at Victor Valley High School, where she was a sophomore. Citing "moral objections to the practice of killing animals for classroom study or for experimentation,"[25] Graham refused to dissect a frog in her biology class. She pleaded with school officials to let her study frog anatomy by using alternatives such as computer simulations, but they denied her request and gave her a low grade for refusing to dissect. She promptly hired an attorney and brought a lawsuit against the school.

After a lengthy battle in court, Graham's case was settled without a trial. The young woman was considered victorious because her grade was raised to an "A," and Victor Valley High school was ordered to pay her legal fees. Her case drew so much national publicity that it inspired an award-winning television special entitled *Frog Girl: The Jenifer Graham Story.* Also, along with support from the Animal Legal Defense Fund, Graham was instrumental in starting a national, toll-free antidissection hotline called *Students Against Dissection*, which logged more than 10,000 calls its first year. The hotline provided callers with information on alternative methods of study, as well as guidance on how to discuss the issue with teachers and other school officials. But for animal rights advocates, Graham's most important victory was paving the way for growing numbers of other students to object to animal dissection in their classes. "Choice-in-dissection" laws were passed in California, which granted students in kindergarten through twelfth grade the right to use alternative methods of study without penalty. Eight other states have passed similar laws, including Florida, Illinois, New York, Pennsylvania, Rhode Island, Oregon, Virginia, and New Jersey, while Louisiana, Maine, and Maryland have adopted student choice policies.

> **"If a student personally believes that it is wrong to harm animals, he or she has a constitutional right to refrain from participating in animal experimentation."**

What Tomorrow Holds

The argument over whether animal experimentation and dissection should be used in the classroom continues today. Those who support it are convinced that education will suffer if students cannot gain hands-on experience with real animals. Animal rights advocates argue that the opposite is true, saying that students can learn just as well, perhaps even better, without having to sacrifice the lives of animals. No matter what happens, animal use in education has seen radical changes over time, and more changes are likely in the years to come.

Should Animal Experimentation Be Used in Education?

"The use of animals by students can be an important component of science education."

—American Association for the Advancement of Science. http://aaas.org.

The American Association for the Advancement of Science is dedicated to advancing science around the world by serving as an educator, leader, spokesperson, and professional association.

"Biology courses should teach students to value and respect all life, but dissection exercises impart the opposite lesson."

—Physicians Committee for Responsible Medicine, "Conscientious Objection in the Classroom." www.pcrm.org.

The PCRM is an organization that opposes what it perceives to be unethical experiments in animals and promotes nonanimal research in medical education.

Bracketed quotes indicate conflicting positions.

* Editor's Note: While the definition of a primary source can be narrowly or broadly defined, for the purposes of Compact Research, a primary source consists of: 1) results of original research presented by an organization or researcher; 2) eyewitness accounts of events, personal experience, or work experience; 3) first-person editorials offering pundits' opinions; 4) government officials presenting political plans and/or policies; 5) representatives of organizations presenting testimony or policy.

"No alternative can substitute for the actual experience of dissection or other use of animals and urges teachers to be aware of the limitations of alternatives."

—National Association of Biology Teachers, quoted in Kansas Junior Academy of Science, *The Use of Animals in Biology Education.* http://webs.wichita.edu.

The National Association of Biology Teachers is an organization that seeks to help educators provide excellence in biology and life science education for students.

"It is imperative that we . . . do everything we can to support those students who are courageous enough to stand up to a frequently oppressive system and to reject the exploitation of [animals]."

—Gary L. Francione and Anna E. Charlton, *Vivisection and Dissection in the Classroom: A Guide to Conscientious Objection.* Jenkintown, PA: American Anti-Vivisection Society, 1992, p. 97.

Francione and Charleton are professors at Rutgers School of Law.

"If we didn't dissect animals, the human race would live entirely based on black magic and superstition."

—Anonymous, quoted in University of Minnesota Academic Health Center, *What High School Students Say About Dissection.* www.ahc.umn.edu.

A student from Hickman High School, Columbia, Missouri.

"Dissecting something that was killed so we could learn about it was unsettling."

—Grace Kendall, quoted in Zinie Chen Sampson, "New Va. Law Requires Schools to Offer Dissection Alternatives," *USA Today*, September 7, 2004. www.usatoday.com.

Kendall, a graduate of Stafford High School in Richmond, Virginia, objects to animal dissection in science education.

"Live demonstrations and experiments involving animals in precollege education are valuable ways to excite children about science."

—American Veterinary Medical Association, *Use of Animals in Precollege Education,* June 2005. www.avma.org.

The American Veterinary Medical Association is the official accrediting body for the 28 schools of veterinary medicine in the United States.

"Each animal cut open and discarded represents not only a life lost, but also a part of a trail of animal abuse and environmental havoc."

—People for the Ethical Treatment of Animals (PETA), "Dissection: Lessons in Cruelty," Media Center. www.peta.org.

PETA is the largest animal rights organization in the world.

"Knowledge, experience, and insights gained through the responsible use of live animals in the classroom and laboratory are unique, invaluable, and irreplaceable elements of a quality education in many basic and clinical disciplines."

—Society for the Study of Reproduction, *Position Statement: Use of Animals in Education,* July 2000. www.ssr.org.

The Society for the Study of Reproduction is an organization of scientists who share a common interest in numerous aspects of science.

"Killing an innocent animal is unethical. The top medical schools produce some of the country's best physicians without ever harming an animal."

—Jeff Tomasini, quoted in Student Doctor Network, *Debated Studies: Animal Labs for Medical Students,* March 19, 2007. http://studentdoctor.net.

Tomasini is a student at the Medical College of Wisconsin.

"Students benefit from the hands-on learning approach that animal laboratories offer."

—Martin Frank, quoted in Student Doctor Network, *Debated Studies: Animal Labs for Medical Students,* March 19, 2007. http://studentdoctor.net.

Frank is executive director of the American Physiological Society in Bethesda, Maryland.

"The dismemberment of dead animals can desensitize students and encourage a view of animals as inanimate objects rather than living beings who once felt pain."

—Cynthia Taylor, "The Case Against Classroom Dissection," *Animal Rights,* About.com. http://animalrights.about.com.

Taylor is an animal rights advocate and environmentalist from Tucson, Arizona.

Facts and Illustrations

Should Animal Experimentation Be Used in Education?

- According to the Humane Society of the United States, an estimated **6 million vertebrae animals** are dissected in American high schools each year, and roughly the same amount of invertebrates, with unknown numbers being dissected in elementary and middle schools.

- Only **13 medical schools in the United States have live animal laboratories:** Johns Hopkins University School of Medicine in Baltimore; Uniformed Services University of the Health Sciences School of Medicine; University of Mississippi Medical Center School of Medicine; St. Louis University School of Medicine; Washington University in St. Louis School of Medicine; New York Medical College; Stony Brook University Health Sciences Center School of Medicine; The Brody School of Medicine at East Carolina University; Case Western Reserve University School of Medicine; Oregon Health & Science University; University of Tennessee College of Medicine; Medical College of Wisconsin; and University of Wisconsin Medical School.

- Nine states have passed **"choice-in-dissection"** laws that grant students the right to use alternative methods of study without penalty, including California, Florida, Illinois, New York, Pennsylvania, Rhode Island, Oregon, Virginia, and New Jersey.

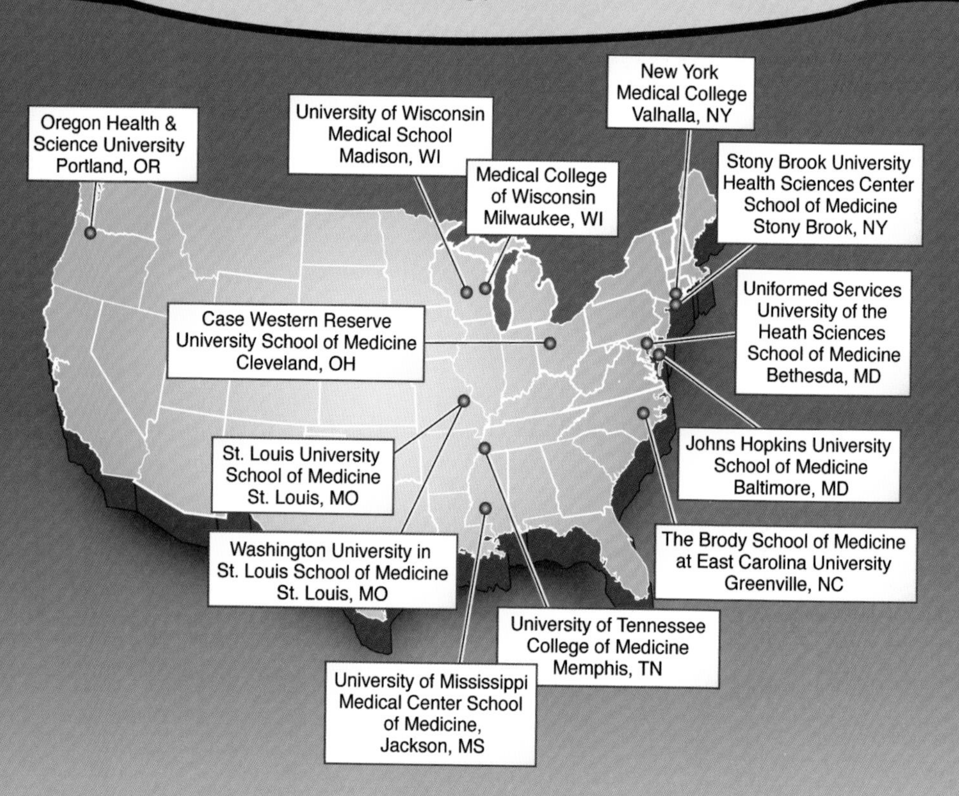

- A March 2004 survey of 749 members of the National Science Teachers Association showed that nearly **80 percent** of educators think dissection is an important part of science learning and **75.2 percent use dissection** in their classrooms.

States with "Choice in Dissection" Laws

As of 2007, nine states have passed laws that allow students from kindergarten through twelfth grade to choose alternatives instead of participating in classroom dissection, while some other states have adopted resolutions or policies that afford similar protection to students. The following map illustrates the states that have such legislation in place.

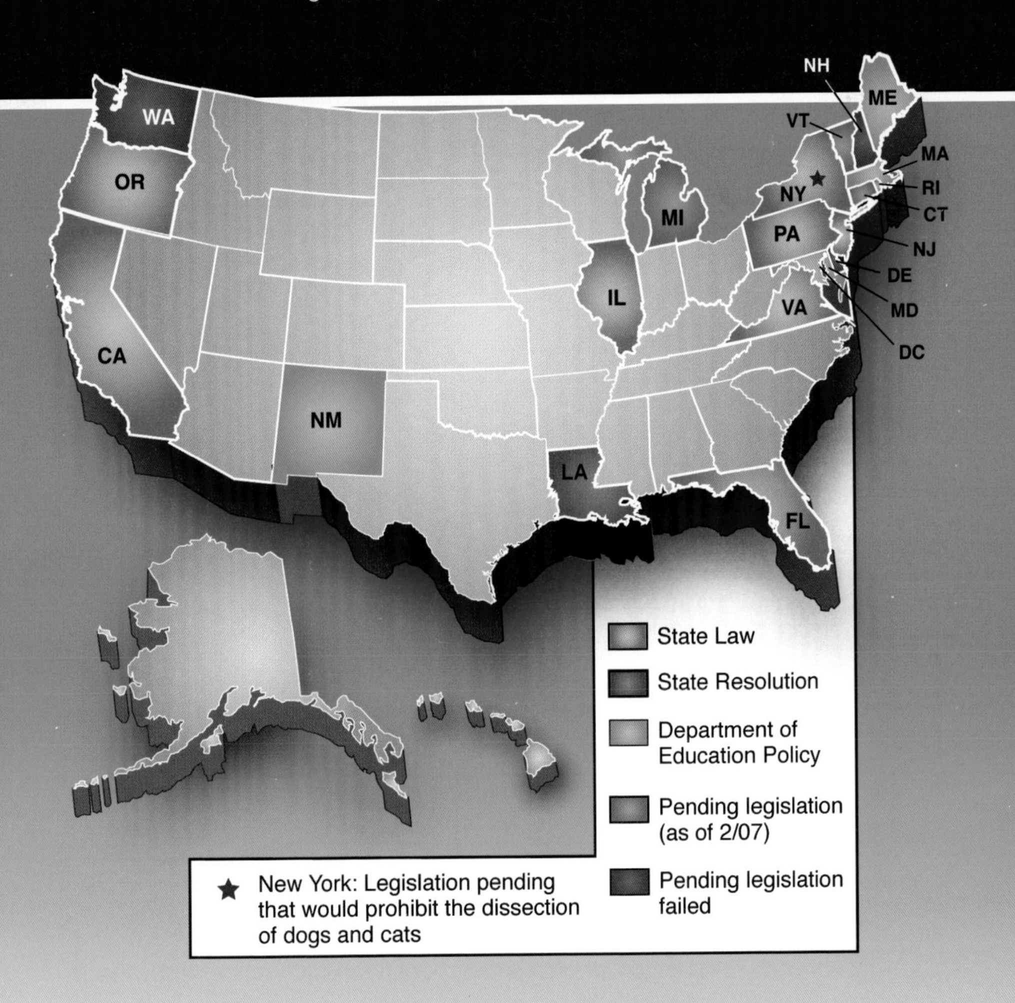

Source: Humane Society of the United States, "Dissection Choice States," February 2007. www.hsus.org.

How Dissection Works

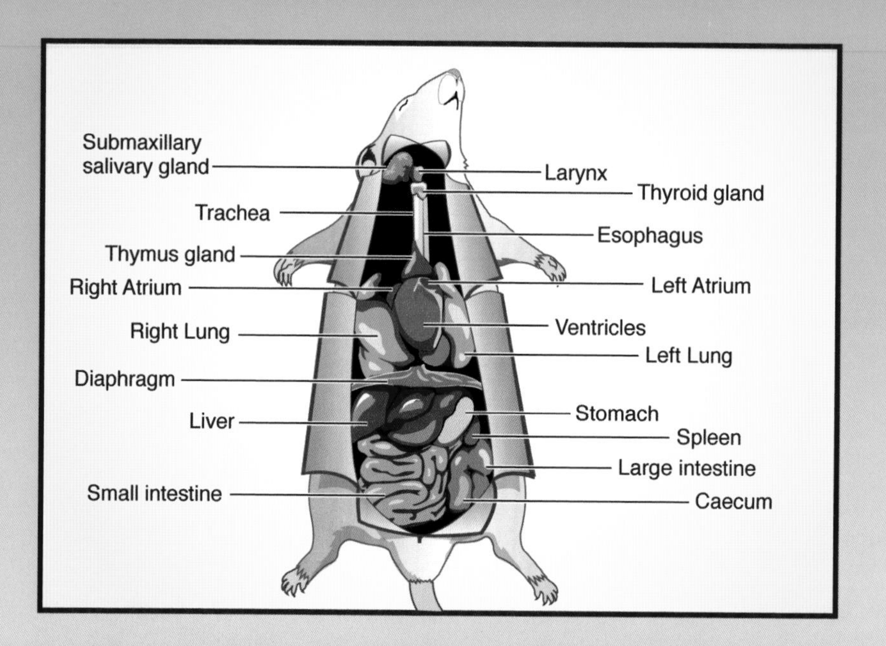

1. The rat is first rinsed off with water to remove any chemicals used in its preservation. Then it is placed in a dissecting pan so students can observe its external body characteristics, such as a hairy coat, whiskers, eyes, ears, and tail.

2. The rat's skin is removed with scissors and forceps to expose the muscles below, and the various muscles are identified and studied.

3. The muscles on the sides of each thigh are pulled away to expose the three leg bones: tibia, fibula, and femur as well as the kneecap.

4. The skin on the neck and face are removed so the salivary glands, lymph glands, and thyroid glands can be studied.

5. The rat's chest is cut open so students can observe its diaphragm, heart, aorta, bronchial tubes, and lungs as well as its entire circulatory system.

6. The abdomen is dissected for exploration of the rat's liver, esophagus, stomach, spleen, pancreas, and intestines.

7. Further dissection allows exploration of the rat's bladder, kidneys, renal arteries and veins, and reproductive organs.

Source: Shannan Muskopf, "Rat Anatomy Dissection Guide," The Biology Corner, September 12, 2007. www.biologycorner.com.

National Science Teachers Association (NSTA) Survey on Dissection in Education

A total of 749 NSTA members participated in a March 2004 survey to gauge opinions of educators on classroom dissection activities. More than 50 percent of the respondents are high school teachers, while the remainder teach at other grade levels.

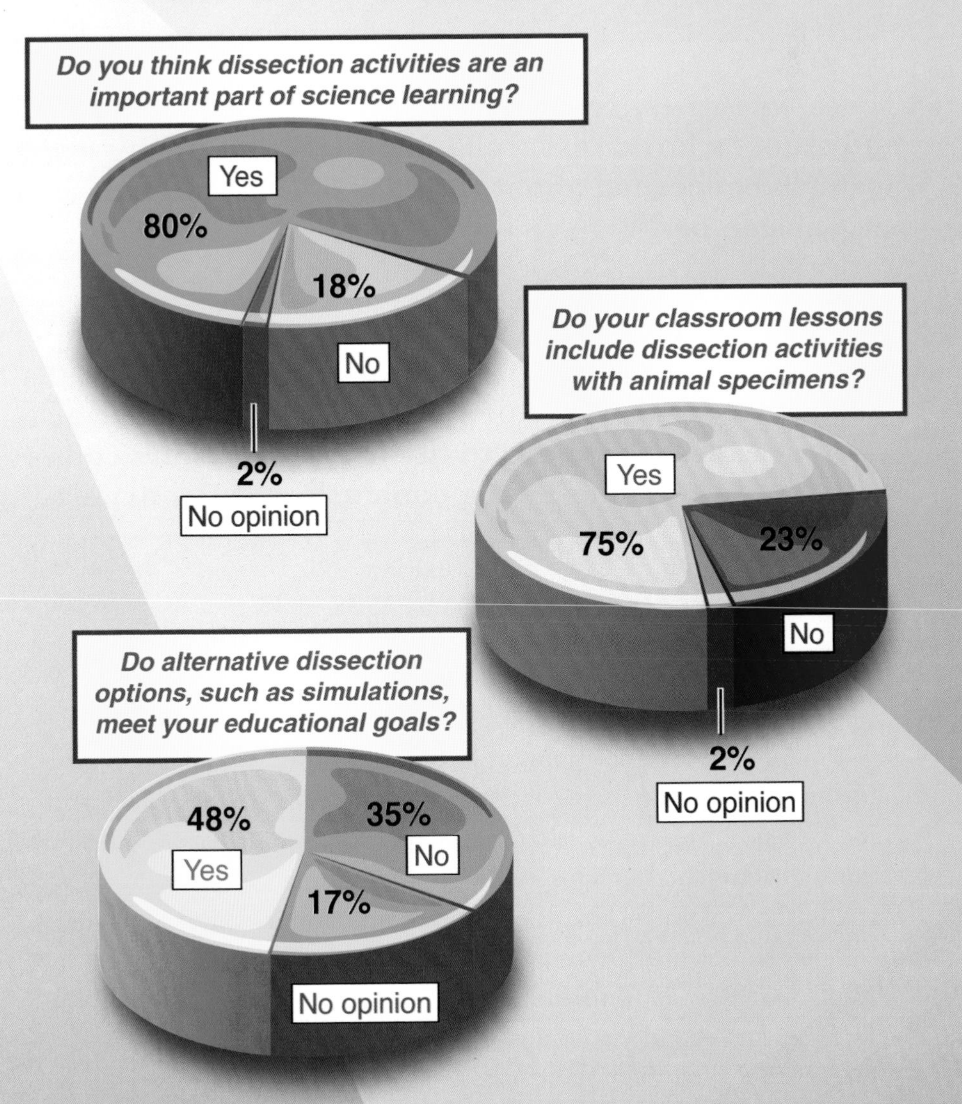

Source: National Science Teachers Association, Survey, "Dissection Survey," March 2004. www.nsta.org.

Are Animal Experiments Conducted Humanely?

"These scientists, veterinarians, physicians, surgeons and others who do research in animal laboratories are as much concerned about the care of their animals as anyone can be."

—Michael E. DeBakey, "Animal Welfare vs. Animal Rights," www.fbresearch.org.

"The general public has virtually no knowledge of what species are used in biomedical research and testing . . . what exactly happens to them over the course of a study, and their fate at the conclusion of the experiment."

—Crystal Schaeffer, "Over 100,000,000 . . . and Counting," 2005. www.aavs.org.

Most scientists who use animal experiments in their laboratories say that every possible effort is made to provide humane care for the creatures. They point at laws such as the AWA and stringent guidelines to which research facilities must adhere. If these laws and guidelines are disregarded, laboratories risk losing valuable funding by various governmental organizations. Also, if animals are not cared for properly, the experiments may be flawed or not produce the kind of quality results that scientists want to achieve.

Officials from Oregon Health & Science University say that the school goes far beyond just abiding by governmental regulations and laws and seeks to exceed these standards of protection. The university's Web site describes the care and compassion that are given to research animals: "Animal caregivers provide food and water, and clean animals' cages or other housing daily. These animals have light, fresh air, treats and toys. Veterinarians regularly check their health status."[26]

Do Laws Adequately Protect Animals?

Opponents of animal experimentation acknowledge the existence of animal welfare laws, but they insist that the laws do not adequately protect animals. One of their concerns is that the AWA is vague and does not specify how a research facility should classify pain. Alan M. Goldberg, professor of toxicology at Johns Hopkins Bloomberg School of Public Health, explains: "A curious observation is that the Animal Welfare Act does not define animal welfare, nor does it define distress, yet the Act is meant to promote animal welfare and reduce or eliminate distress."[27] Goldberg adds, though, that there are "five freedoms" researchers are expected to adhere to with regard to humane animal care: the freedom to have access to fresh water at all times; freedom to have enough nutritious food; freedom from pain and distress; freedom from anxiety and fear; and freedom to be able to engage in normal behavior, such as grooming, foraging, and hiding.

> **"Opponents of animal experimentation acknowledge the existence of animal welfare laws, but they insist that the laws do not adequately protect animals."**

Because of how the AWA is worded, rats, mice, and birds—which comprise from 90 to 95 percent of all animals used for research—are not protected, even in terms of minimal standards of care. That was not always the case, however. Although the original Laboratory Animal Welfare Act in 1966 excluded mice, rats, and birds, it was amended in 1970 to cover all warm-blooded animals used in research, testing, and experimentation and also required that research facilities provide pain relief. Then in 1971 the AWA was amended again to specifically exclude rats, mice, and birds. Since then it has been amended several more times, including in 2002 when the Farm Security and Rural Investment Act was passed, which again made it clear that rats, mice, and birds, were excluded. Senator Robert Dole, who had worked on trying to expand the AWA's protection, wrote about his frustration in a letter to the Alternatives Research and Development Foundation: "As someone deeply involved with the process of revising and expanding the provisions of the AWA, I assure you that the AWA was meant to include birds, mice, and rats. When Congress stated that the AWA applied to 'all warm-blooded animals,' we certainly did not intend to exclude 95 percent of the animals used in biomedical research laboratories."[28] Although mice, rats, aquatic animals, amphibians, and birds are offered some protection in the National Institutes of Health's *Guide for Laboratory Animal Facilities and Care*, the guide does not have the force of law or regulation, and facilities can deviate from it if they believe they have valid scientific reasons for doing so.

AWA Violations

A September 2005 report by the USDA's Office of Inspector General (OIG) found that many research laboratories were violating the law. From 2002 through 2004 the number of research facilities that were cited for violations of the AWA had steadily increased, but when the violations were reported, offenders got off easily. In order to avoid creating hostility with research scientists, and in an effort to reach amicable agreements without ending up in court, USDA officials gave the violators an automatic 75 percent discount off their fines. One violation discovered by OIG inspectors was that some laboratories were providing erroneous reporting. A research facility in New York, for example, reported that it housed three primates, but USDA inspectors found that 42 primates were actually at the lab. A laboratory in California that had been approved for experiments on 80 rabbits was later found to have used more than 1,000 rabbits.

Also included in the inspectors' report was their discovery that many facilities were neglectful of animals or conducting research that caused pain or distress without providing anesthetics or other drugs to help ease the creatures' suffering. For instance, a research facility in Illinois failed to provide proper veterinary care, which resulted in the death of a primate and a pig. At a laboratory in California, researchers failed to give any pain medication to a primate after the creature had just undergone surgery in which holes were drilled into its skull. At Emory University, two monkeys were duct taped to restraint chairs so blood could be drawn. At the Southern Research Institute, dogs were starved for two weeks, and one of them lost nearly 40 percent of its body weight.

During the summer of 2004 the USDA issued a formal complaint against the University of California at San Francisco, which stated that "the gravity of the [UCSF] violations is great"[29] and cited 60 incidents that occurred in a two-year period. The complaint said the research facility failed to provide adequate veterinary care and sanitary conditions, neglected to administer anesthetics, did not properly monitor animals following surgery, and deprived animals of adequate water. A lamb that had undergone surgery was not monitored properly, and when the creature was discovered by an attendant, it was clearly suffering, frothing at the mouth, and having difficulty breathing. A sheep was not given anesthesia before surgery, nor was she given pain relief after surgery. A female marmoset (squirrel monkey), which only weighed 14 ounces (400 grams), gave birth 7 times in 40 months to 14 babies, 6 of which died. According to USDA inspectors, in July 2002 when the marmoset had her last litter, she weighed just 10 ounces (283 grams). Laboratory attendants separated her and one of her babies from the group cage, and when the creatures were discovered, the mother was thin and clearly depressed, and the baby was so lethargic and nonresponsive that it was euthanized. The mother was later euthanized as well.

> **"A September 2005 report by the USDA's Office of Inspector General (OIG) found that many research laboratories were violating the law."**

"Humane Animal Care and Good Science Go Hand in Hand"

Many laboratories do a good job of ensuring that animals receive proper veterinary care and are treated humanely. They abide by the laws, often exceeding the standards of care that the government requires and treating animals with respect. The University of Michigan is home to more than 142,000 research animals, including cats, dogs, chickens, primates, sheep, frogs, guinea pigs, rats, and mice. Howard G. Rush, the director of the school's Laboratory of Animal Medicine, says that all the people who work there love animals and take excellent care of them. Technicians inspect each animal at least once a day, and more often if they have just had surgery. If an animal is in pain or shows signs of infection, sickness, or injury, a veterinarian or veterinary technician provides immediate medical care. Rush also says that the animals are given anesthesia during surgery or any invasive procedure, and they are given the same pain-killing medications used in veterinary clinics or human hospitals. Some of the U of M's research animals live out their natural life spans and die of old age. This was the case with the school's most famous mouse, Yoda. In 2004 Yoda died at the age of 4 years and 12 days, which is the equivalent of 136 human years.

> **"Many laboratories do a good job of ensuring that animals receive proper veterinary care and are treated humanely."**

The Southwest Foundation for Biomedical Research (SFBR), located in San Antonio, Texas, says it is wholly committed to providing exceptional care for animals used in research. The facility has been commended for achieving high standards of care and for exemplary good health and well-being of its animals. The SFBR Primate Research Center's staff includes seven full-time veterinarians and more than 130 veterinary technicians and other animal care professionals, all of whom are charged with providing humane care and treatment. The organization explains: "The care of animals at SFBR is guided by two principles: first, that animals deserve high-quality care and state-of-the-art preventive medicine; and second, that high-quality scientific data can be derived only from animals that are treated humanely and provided with proper care. . . . Humane animal care and good science go hand in hand."[30]

Can Animal Suffering Be "Humane"?

Even if animals are well cared for in laboratories, animal experimentation opponents believe that by its very nature, the practice is inhumane. While there are standards of care for many research animals, there is no experiment that is specifically prohibited by U.S. law. As a result, tests such as LD50 and Draize are legal, no matter what animals they are used on or how much the animals may suffer. Rachel Menge and Michelle Thew of the Coalition for Consumer Information on Cosmetics describe what creatures must endure when they are used in LD50 laboratory tests that gauge the toxicity of various chemicals: "Death comes only after the animal experiences the grisly effects of the poisoning. Sometimes the volume alone of the substance the animals are forced to ingest actually kills them. Other times the test measures the experienced ghastly effects of the chemical destroying the animal's internal organs until death."[31] Depending on the research being done, scientists use from 50 to 100 animals for each LD50 experiment, and at least half of the creatures die by the time the tests are complete.

The Humane Society of the United States has been active in the fight against LD50 experiments being used to test Botox Cosmetic, a drug produced by the pharmaceutical company Allergan. Although Botox injections are sometimes used for therapeutic reasons, the drug has become a popular treatment for wrinkles, such as crows' feet around the eyes. Botox, which contains the deadly botulinum toxin, works by temporarily paralyzing the muscles in a patient's face. To determine the proper strength of each batch, Allergan researchers must test it on animals before it is released to doctors and dermatologists. The HSUS describes the effects of these tests:

"Even if animals are well cared for in laboratories, animal experimentation opponents believe that by its very nature, the practice is inhumane."

> Imagine that you are one of the animals unfortunate enough to be used in assessing the potency of new batches of Botox Cosmetic. . . . First, the toxic substance or the full product is injected into your stomach. Then, as it courses through your bloodstream, the toxin causes nausea and then brings on a wave of muscle paralysis that spreads throughout your body. Finally, over the course of the three- to four-day test, you suffocate to death.[32]

Are Some Animal Experiments Bad Science?

When discussing whether animal research is humane, those who are opposed to such practices say that many experiments are unnecessary and in some cases represent unsound science. Jonathan Balcombe of the Humane Society of the United States cites such a study, which was performed by a team of researchers and described in a March 2006 issue of *Science.* The experiments were designed to study the effects of clinical depression in an effort to develop a treatment for humans. The researchers were trying to determine the ability of a protein called p11 to fight depression, so some of the mice used in the experiments were genetically altered to lack the protein. Part of the research involved "tail suspension," during which adhesive tape was attached to the mice's tails and the creatures were dangled upside down for six minutes. In other experiments mice and rats were given electric shocks to the head through special electrode clips attached to their ears. According to Balcombe, the tests were inhumane as well as unscientific. "The treatment of the mice is depressing in itself . . . If we wanted to see if something caused headaches in people, we wouldn't want them to be suffering headaches to begin with. It is common for animals in laboratories to suffer from baseline depression. How can we adequately assess 'new' depression in an already depressed animal?"[33]

"Part of the research involved "tail suspension," during which adhesive tape was attached to the mice's tails and the creatures were dangled upside down for six minutes."

Whether animals used in research are treated humanely is often a matter of personal opinion. Some laboratories insist that they go far beyond what is required by law and provide animals with exceptional standards of care, while others are exempt from government supervision altogether. Still others have reported erroneous information or have been found in violation of the law. Animal experimentation opponents believe that existing laws do not adequately protect animals at all—and even when those laws are adhered to, the experiments themselves are inherently wrong.

Primary Source Quotes*

Are Animal Experiments Conducted Humanely?

“Humane animal care is a basic necessity in medical research, not only for ethical reasons, but because scientists cannot obtain valid results from mistreated animals.”

—Michigan Society for Medical Research, “Other Information: Fact vs. Myth.” www.mismr.org.

The Michigan Society for Medical Research is a scientific educational organization that promotes understanding of biomedical research and testing, including the humane use of animals.

“Animals in laboratories are routinely subjected to painful procedures and are usually killed afterward. Routine caging, isolation, handling, and even the laboratory environment itself are extremely stressful to animals.”

—Physicians Committee for Responsible Medicine, “Frequently Asked Questions About Animal Experimentation Issues.” www.pcrm.org.

The PCRM is an organization that opposes unethical experiments in animals and promotes nonanimal research in medical education.

Bracketed quotes indicate conflicting positions.

* Editor’s Note: While the definition of a primary source can be narrowly or broadly defined, for the purposes of Compact Research, a primary source consists of: 1) results of original research presented by an organization or researcher; 2) eyewitness accounts of events, personal experience, or work experience; 3) first-person editorials offering pundits’ opinions; 4) government officials presenting political plans and/or policies; 5) representatives of organizations presenting testimony or policy.

"While we all wish that we could advance health and medical research without involving animals, we know that some of science's greatest advances in saving humans could only have come from this research."

—Julie Louise Gerberding, Centers for Disease Control and Prevention, November 16, 2006. www.cdc.gov.

Gerberding is director of the CDC as well as a professor of medicine at Emory University and the University of California at San Francisco.

"I witnessed several animals suffer from behavioral and health issues [that] were a direct result of the care they were not receiving from staff."

—Cheri Stevens, "Cheri Stevens Statement: California National Primate Research Center (CNPRC)," Stop Animal Exploitation NOW. www.all-creatures.org.

Stevens is a former employee of the California National Primate Research Center at the University of California at Davis.

"Animal activists prey on the emotions of pet owners. They falsely claim that pets are stolen and sold to medical research facilities and suppliers of animals for scientific research."

—Carolina Biological Supply Company, "Dissection Fact Sheet." www.carolina.com.

Carolina Biological is the largest animal supply company in the United States.

"The general public and scientists alike are troubled by the fact that animals are suffering from pain, fear, and anxiety in the course of biomedical experimentation."

—Martin Stephens, quoted in "Poll Shows Americans Disapprove of Animal Research When It Causes the Animals to Suffer," Humane Society of the United States, November 14, 2001. www.hsus.org.

Stephens is vice president of animal research issues for the Humane Society of the United States.

"I do feel bad about using rats but I can sleep at night—we use the fewest possible and try our best to ensure they suffer as little as possible."

—Anonymous, quoted in Emma Marris, "Grey Matters," *Nature,* December 14, 2006.

Quote by a researcher from the United Kingdom who did not want to give her name.

"In the case of experimentation, it is often argued that the potential benefits to humans justify keeping animals in unnatural and highly confined conditions and causing them pain, suffering and distress. We disagree with this anthropocentric viewpoint."

—Gill Langley, "Next of Kin: A Report on the Use of Primates in Experiments," British Union for the Abolition of Vivisection, June 2006.

Langley has a PhD in neurochemistry and is a scientific consultant to the British Union for the Abolition of Vivisection and other animal protection organizations.

“Laboratory animal veterinarians and other animal caregivers have a legal and moral obligation to alleviate pain and distress in laboratory animals.”

—American Association for Laboratory Animal Science, “Position Statements.” www.aalas.org.

The American Association for Laboratory Animal Science is an organization devoted to the humane care and use of laboratory animals in scientific research.

“Animal experimentation . . . is arguably the most severe form of systematic violence in the modern world.”

—Uncaged, “Animal Experimentation: The Facts.” www.uncaged.co.uk.

Uncaged is an international animal rights organization located in the United Kingdom.

“Good science and good animal care are inseparable.”

—National Institute of Environmental Health Sciences (NIEHS), *Respect for Life: Myths and Facts About Laboratory Animals*, June 2000. www.niehs.nih.gov.

The NIEHS, which is affiliated with the National Institutes of Health, focuses on research that leads to a greater understanding of human health and disease.

“Animals have no voice and no choice; we have to speak out for them.”

—Julie Roberts, quoted in BBC News, “Viewpoint: Animal Rights,” September 7, 2004. http://news.bbc.co.uk.

Roberts is a noted campaigner for animal rights in the United Kingdom.

Facts and Illustrations

Are Animal Experiments Conducted Humanely?

- The **Animal Welfare Act (AWA)** was originally passed in 1966, and in 1970 it was amended to ensure greater protection of warm-blooded animals used in research.

- In 2002 the AWA was weakened by passage of the Farm Security and Rural Investment Act, which modified the AWA's definition of animals **to exclude rats, mice, and birds used in research**.

- According to the USDA, nearly **85,000 animals** used in research during 2005 experienced pain and were given no anesthetics or drugs to alleviate suffering.

- U.S. animal protection laws **do not prohibit any experiment,** including LD50 and Draize tests.

- While U.S. military and other government laboratories are expected to abide by the AWA, they **are not under the authority** of the USDA.

- The European parliament has voted to **ban the sale of animal-tested cosmetic products** in Europe and outline the use of animals for cosmetic testing by the year 2009.

Pain, Suffering, and Distress Is Minimized

According to reports published in 2006 by the USDA, only a small percentage of the animals used in research during 2005 experienced significant pain or distress.* Over 50 percent of the animals experienced no pain, so no drugs were necessary; 36 percent experienced pain and were given drugs to alleviate suffering; and 7.2 percent experienced pain but were given no anesthesia or painkillers, either because the experiments themselves focused on studying pain, or because drugs would have interfered with the test results.

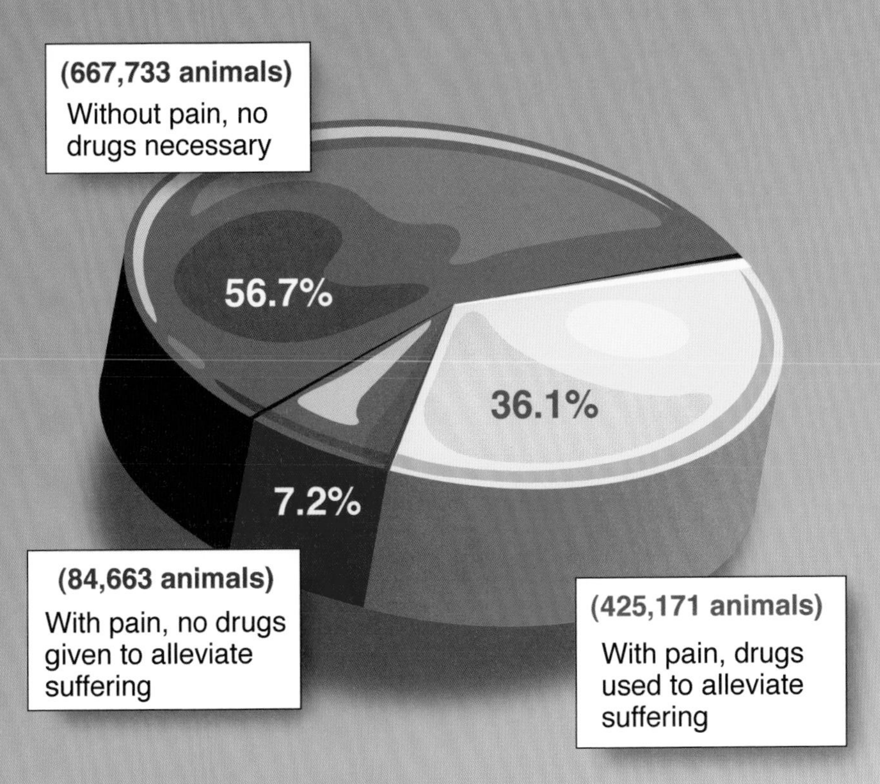

* There are no data that show pain and suffering by mice, rats, birds, or other animals that are not included in the USDA's reporting.

Source: United States Department of Agriculture, Animal, and Plant Health Inspection Service, "FY 2005 AWA Inspections." www.aphis.usda.gov.

The Draize Test

The Draize Test is commonly used for testing chemicals, household cleaners, cosmetics, toiletries, and a number of products to see what damage could be caused to people.

This photo shows rabbits confined in head-holding devices so that chemicals or other products can be dripped into their eyes to see how much damage is caused.

In this photo, guinea pigs have been used to test chemicals contained in cosmetics to gauge how badly their skin is damaged.

Source: Brian Gunn, International Association Against Painful Experiments on Animals. www.animalexperimentspictures.com.

Top 20 Violators of the Animal Welfare Act During 2005

This ranking is based on inspections conducted by the Animal and Plant Health Inspection Service (APHIS), a division of the USDA that inspects facilities that use animals in research.

Facility	Total Violations for a 9-month period	Number of repeat violations
University of Pennsylvania	77	15
Harvard University	32	2
Covance Laboratories (Pennsylvania)	27	
Charles River Laboratories	22	1
University of Wisconsin	20	7
Boehringer Ingelheim Pharmaceuticals	19	1
Covance Laboratories (Virginia)	15	
University of Connecticut	15	
Vanderbilt University	13	
Emory University	13	1
Boston University	11	7
McLean Hospital	11	4
Schepens Eye Research Institute	11	1
Brigham & Women's Hospital	10	
Massachusetts Institute of Technology (MIT)	10	4
Massachusetts General Hospital	9	4
Stanford University Veterinary Service Center	9	1
Franklin & Marshall College	9	
University of New Jersey Medical & Dental School	8	
Merck Research Laboratory	8	
Shin Nippon Biomedical Laboratories (Washington)	8	
Totals	**357**	**48**

Source: Stop Animal Exploitation NOW! (SAEN), "The Top 20 Violators of the Animal Welfare Act," February 23, 2007. www.all-creatures.org.

Can Alternative Methods Eliminate the Need for Animal Experimentation?

"If there were good alternatives to animals that worked better or as well, for less money and hassle, scientists would use them. We can be stubborn but we are not totally bananas."

—Stuart Derbyshire, Edinburgh Book Festival, 2002.

"The animal research industry is fixated on the dangerous myth that profit-driven animal research is necessary . . . it clearly is not the best 'fix' for understanding or treating human health and disease."

—Theodora Capaldo, New England Anti-Vivisection Society press release, 2002.

In 1959 British scientists William Russell and Rex Burch urged researchers all over the world to minimize the use of animals in their laboratories. When animals were absolutely necessary for research, Russell and Burch urged that the creatures' pain and stress be minimized. They gave a detailed explanation in a book entitled *The Principles of Humane Experimental Technique*, in which they introduced the three Rs: replacement, reduction, and refinement. In describing the concept of replacement, Russell and Burch stated that whenever possible, living animals should be replaced with alternative methods, and researchers were encouraged to actively search for methods that did not require the use of animals. If it was determined that animals were absolutely necessary, the number of creatures used should be reduced to the smallest number possible. The third "R," refinement, pertained to improving the lives of animals used in research, such as ensuring that they were always treated humanely; had adequate fresh air, water, and food; and were provided with living conditions that were as pleasant as possible. Russell and Burch were convinced that even with replacing, reducing, and refining animal use, it would still be wholly possible for crucial scientific and medical progress to be made.

Many scientists today strive to achieve the goals of the three Rs, believing that the concept represents sound science combined with humane, respectful treatment of animals. But many animal advocacy groups, such as the Association of Veterinarians for Animal Rights (AVAR), do not believe that the three Rs go far enough in protecting animals. The AVAR explains this on its Web site:

"Many scientists today strive to achieve the goals of the three Rs, believing that the concept represents sound science combined with humane, respectful treatment of animals."

> Whereas we applaud the industry's responsiveness and appropriation of money for alternatives, we believe the current trend misses the point. We . . . want the industry to completely do away with the use of . . . animals or their bodily parts on the basis that it is unnecessary and inhumane. There are numerous products presently available that have not been tested on any . . . animals, do not contain ingredients from . . . animals and are reasonable alternatives to what the industry presently offers."[34]

Replacing Animals in Research

Over the years many scientific advances have led to animals being replaced with alternate research techniques. Actually, "replacement" falls into two categories: *Absolute replacement* occurs when no animals are used at all, while *relative replacement* involves only the use of animal cells or tissue. Understanding the difference between the two categories can be confusing, as it is somewhat deceptive. Even though research is not performed on live animals in relative replacement methods, the cells or tissue often comes from animals that were killed specifically for that research.

Absolute replacement led to the complete elimination of one type of animal research known as the "rabbit test." In the past, such tests were a common way for women to find out if they were pregnant. This began in the late 1920s, when researchers discovered that a hormone known as HCG was present in a woman's urine shortly after a fertilized egg implanted itself on the wall of her uterus. They injected female rabbits with human urine that contained HCG, and after killing the creatures to remove and examine their ovaries, they found that the ovaries had undergone distinct changes. Thus, the rabbit test was born. Today such testing is obsolete, having been replaced with over-the-counter pregnancy test kits that allow women to test HCG levels in their own urine.

In a growing number of experiments, animals are being replaced with tests that are known as in vitro, which means "in glass." In vitro research experiments are conducted in test tubes and are designed to determine the effects of drugs or vaccines, or the biological processes of viruses, on cells and tissue rather than in the bodies of live animals. In vitro tests have also been hailed as beneficial for more reasons than just eliminating the need for animals, as an article in *Scientific American* explains: "Hormones or vaccines manufactured in cell cultures are also purer than those made in vivo (that is, in the animals themselves), so each batch need not be tested as before for safety and efficacy."[35] Many people believe that the potential for in vitro technology is enormous, and its use could save the lives of hundreds of thousands of research animals every year.

"Many people believe that the potential for in vitro technology is enormous, and its use could save the lives of hundreds of thousands of research animals every year."

One type of in vitro test uses a substance called Corrositex, which is made of a protein membrane that serves as a kind of synthetic skin.

Corrositex was designed to test chemicals, acids, and other caustic substances for corrosiveness, or their ability to burn or destroy skin. Using a color-change process, the test can detect even the most minute chemical changes in cells and measure the rate at which a substance is capable of causing damage. A similar in vitro test is known as Skintex, which uses pumpkin rind to mimic the reaction of an alien substance on human skin, thereby gauging the substance's harmfulness—again, in the test tube. Another research product, known as Testskin, uses human skin that is grown in a sterile plastic bag, and it can also be used to measure the irritant potential of numerous substances. An in vitro test designed to measure potential eye irritation is known as Eyetex. It uses a compound of proteins, one of which is derived from the jack bean, to mimic the reaction of the human cornea to a foreign substance. When chemicals are added to the proteins, the degree of irritation can be measured by the levels of cloudiness.

Reducing Animals in Research

Even if research cannot be done without using animals, many scientists are seeking ways to reduce the number of creatures they use in their laboratories. According to the Humane Society of the United States, laboratories throughout Europe, Japan, and North America report that animal usage has dropped by about half of what it used to be. As technology advances, more and more products and tests are available to help scientists reduce the amount of animals needed for research. At Oregon Health & Science University, for example, researchers have found ways of cutting down on the number of animals that are needed for medical training. In the past, pigs were used to teach medical residents how to perform lifesaving techniques. Today, specially designed humanlike manikins are used for this training, rather than live animals.

Veterinary surgeon Christian Schnell has markedly reduced the number of animals he uses for research—and at the same time, he has lessened the amount of stress the creatures go through in his experiments. In the late 1980s Schnell studied blood pressure data from marmosets by physically restraining the creatures, catheterizing their arteries, and forcibly giving them blood pressure medication. After the experiments were completed, the animals were killed. Schnell was dismayed at the suffering the marmosets went through, as well as having to euthanize them, so he developed a new method of study. He implanted a tiny sensor in their abdominal cavities, which he could use to monitor their heart rates and blood pressure as the creatures went about their normal routines. Because he switched to the new technique, Schnell's marmoset use decreased about 90 percent. He is also convinced that his results became more accurate because the research animals were not stressed as they had been with his old methods.

Alternatives to Using Animals in Education

In the same way that technology has helped scientists replace or reduce animals in their research, numerous products have been designed especially for educational institutions. One product available to medical schools is a computerized manikin known as TraumaMan, which is used in advanced trauma life support (ATLS) courses for aspiring surgeons and doctors who work in emergency rooms. Typically, live pigs, goats, dogs, or sheep have been used in ATLS classes, and once the course was over, the animals were killed. TraumaMan helps students master their emergency surgical skills without experimenting on live animals. Nell Boyce of NPR describes the high-tech simulator: "He's a headless, legless, armless torso with nipples and a belly button. His ribs bulge beneath pink, rubbery skin. The chest rises and falls with each mechanical breath. But cut TraumaMan, and does he not bleed? Well, yes—or at least, he delivers a 'blood flow response' when cut by a scalpel."[36]

"Because he switched to the new technique, Schnell's marmoset use decreased about 90 percent."

In college psychology classrooms, *Sniffy the Virtual Rat* allows students to study the behavior of a rat and how that behavior can be conditioned, without using live creatures. The computer-simulated Sniffy mimics a real rat, pacing around its barren cage with water available but no visible food. The objective is for students to "train" the rat to press a bar above a dispenser, whereby he will be rewarded by receiving virtual food pellets.

Technology also makes it possible for secondary schools to use high-tech computer programs in place of animals. Virtual reality software, for instance, can replace traditional dissection in biology classrooms. One device called the *Virtual Frog Dissection Kit* allows students to learn about the anatomical structures of frogs by "dissecting" one on the screen, while *Dissection Works* offers students interactive dissection simulations of a frog, crayfish, perch, fetal pig, and cat. According to Nancy Harrison, a pathologist from San Diego, this type of technology has immense potential in education. She explains: "Computerized dissection alternatives have grown so sophisticated they now surpass traditional wet dissections in many ways. Numerous studies published in the literature of the education profession demonstrate same or better academic performance by students who study alternatives."[37]

In the same way that technology has helped scientists replace or reduce animals in their research, numerous products have been designed especially for educational institutions.

Despite the numerous programs that are designed to replace traditional dissection in classrooms, many educators and students still believe that nothing can substitute for dissecting a real creature. Becky Baumgartner shares her views:

> Most people who oppose dissections are in favor of online guides and diagrams as an alternative to actual dissection. While I'm sure that these programs are useful and effective in learning to identify the parts, you cannot fully grasp the magnificence of life without examining a real animal up close. For example, imagine a great novel—a classic, a true literary masterpiece. You can stare at the cover all you like; you can read the cliff notes that somebody else wrote; but you will never know just how great and wonderful it is until you delve into it yourself."[38]

Animal Experimentation in the Future

With such radical diversity of opinion on the subject of animal experimentation, no one can say what its role will be in the future of research, education, and product testing. Technology, combined with people's growing concern for the welfare of animals, has led to significant changes in how laboratories conduct research, and animal rights advocates are hopeful that these factors will eventually bring an end to the use of animals in research altogether. Another factor is fear: Violence by extreme animal rights groups has steadily increased since the 1980s, which has caused animal researchers to become much more secretive about their work out of fear for their lives and the lives of their families. According to a February 2006 report by the Foundation for Biomedical Research, illegal acts by animal extremists increased more than 1,000 percent from 1980 through 2003, and many of the incidents involved acts of arson or bombing. In 2005 a man connected with the ALF was arrested for bombing the home of an executive with the pharmaceutical research company GlaxoSmithKline, which is located in the UK. Known as the ALF's "leading bomber," the man was sentenced to prison for 12 years. In the United States a fire that was set by animal rights extremists in August 2003 destroyed a five-story condominium in San Diego, causing $50 million in damage. Because of the risks involved, some laboratories have been forced to stop using animals in their research. Others have begun outsourcing their work to China, where animal welfare laws are not stringent and animal experimentation and testing are more accepted than in the United States.

> **"With such radical diversity of opinion on the subject of animal experimentation, no one can say what its role will be in the future of research, education, and product testing."**

Even with the expanding acceptance of the three Rs philosophy, along with society's growing concern about how animals are treated, many scientists remain convinced that animal experimentation will never be completely eliminated. As the CDC explains: "Whenever possible, new methodologies are used to avoid the use of animals, reduce the number of animals required to attain the intended results, or lessen the impact on the animals. . . . However, the search for an alternative will, at times, be a dead end. In simple terms, animal research will always be required."[39]

Can Alternative Methods Eliminate the Need for Animal Experimentation?

"I want to see the total end of animal experimentation, but I am not stupid enough to think that is going to happen overnight."

—Les Ward, quoted in Emma Marris, "Grey Matters," *Nature,* December 2006.

Ward is director of the animal rights group Advocates for Animals.

"There is a totally hard-headed and rational case to be made for saying that animal experimentation has been a scientific and medical disaster."

—Jerome Burne, "Animal Testing Is a Disaster," *Guardian,* May 24, 2001.

Burne is a science and medical journalist from the United Kingdom.

Bracketed quotes indicate conflicting positions.

* Editor's Note: While the definition of a primary source can be narrowly or broadly defined, for the purposes of Compact Research, a primary source consists of: 1) results of original research presented by an organization or researcher; 2) eyewitness accounts of events, personal experience, or work experience; 3) first-person editorials offering pundits' opinions; 4) government officials presenting political plans and/or policies; 5) representatives of organizations presenting testimony or policy.

"Unfortunately there is no substitute to testing on live animals. . . . If we stopped animal testing, drug development would stop short."

—Glenn Rice, quoted in *Forbes,* "Comparative Advantage," November 13, 2006.

Rice is chief executive of Bridge Pharmaceuticals, a drug developer for the pharmaceutical and biotechnology industries.

"It is only out of sheer habit or ease that scientists continue to inflict pain on animals when, in fact, alternatives exist. And, where alternatives don't exist, the moral task of science is to discover them."

—Claire Andre and Manuel Velasquez, "Of Cures and Creatures Great and Small," *Issues in Ethics,* Spring 1988.

Andre and Velasquez are with the Markkula Center for Applied Ethics at Santa Clara University.

"Although much research is conducted in the test tube and by computer analysis, ultimately a potential treatment, cure or preventive strategy frequently must be tested on a complex living organism."

—Oregon Health & Science University, "Why Use Animals in Research?" http://onprc.ohsu.edu.

OHSU is a health and research university in Beaverton, Oregon, and the site of the Oregon National Primate Research Center.

"Given the choice of apathy or someone liberating mink, burning down a research torture laboratory, or killing a vivisectionist or other DIRECT murderer of animals, I will choose the aforesaid actions over apathy any day of the week."

—Gary Yourofsky, quoted in Jacob Laksin, "Animal Rights Extremism Meets Academia," *Front Page Magazine,* April 19, 2007.

Yourofsky is active in the extreme animal rights group Animal Liberation Front (ALF).

"Are we really so fed up with humanity that we would sacrifice ourselves and our children to save the monkeys?"

—Stuart Derbyshire, animal experimentation speech, Edinburgh Book Festival, August 19, 2002.

Derbyshire is a senior lecturer at the United Kingdom's University of Birmingham.

"We need a new mind set: recognizing that the use of sentient beings for this purpose is essentially unethical we must set the human brain, linked with the heart, to find new ways forward without the use of any of these animals."

—Jane Goodall, "Foreword," in Gill Langley, *Next of Kin: A Report on the Use of Primates in Experiments.* London: British Union for the Abolition of Vivisection, June 2006.

Goodall is a noted authority on primates, as well as an author and the founder of the Jane Goodall Institute.

“The goal of the ‘animal rights’ movement is not to stop sadistic animal torturers; it is to sacrifice human well-being for the sake of animals.”

—Alex Epstein, “Animal Rights Movement’s Cruelty to Humans,” *Capitalism Magazine,* August 15, 2005.

Epstein is a writer for the Ayn Rand Institute in Irvine, California.

“If we are truly prepared to dismiss consciousness in mammals, it seems only a step away from dismissing it in very young children and infants.”

—Douglas Watt, quoted in Gill Langley, *Next of Kin: A Report on the Use of Primates in Experiments.* London: British Union for the Abolition of Vivisection, June 2006.

Watt is the director of neuropsychology at the Boston University School Medicine.

“I would like to see a day when animals are no longer needed to find cures for illness and disease, but that day is not yet here.”

—Jo Tanner, “Standing Up for Animal Research,” *Chemistry and Industry,* May 1, 2006.

Tanner is the chief executive of the Coalition of Medical Progress.

"Even if Animal Research Resulted in a Cure for AIDS, We'd Be Against It."

—Ingrid Newkirk, quoted in Michael Fumento, "PETA: People Enabling Terrorist Atrocities," *Tech Central Station,* June 15, 2005.

Newkirk is the cofounder and president of People for the Ethical Treatment of Animals.

Facts and Illustrations

Can Alternative Methods Eliminate the Need for Animal Experimentation?

- A 2005 survey commissioned by the Physicians Committee for Responsible Medicine showed that **71 percent** of people who donate to health charities want their donations to be used for innovative nonanimal research rather than animal experiments.

- According to the Humane Society of the United States, laboratories throughout Europe, Japan, and North America report that animal usage has dropped by about **50 percent** of what it used to be.

- **New laws are changing animal experimentation:** In New York, it is now illegal to use animals to test for irritants in cosmetics, except where such tests are required by federal law, while in Arizona and New Jersey, animals cannot be used for product testing if suitable alternative tests exist.

- In 2005 the Humane Society of the United States **won a lawsuit** against the USDA, thereby forcing the agency to post the annual reports of registered research facilities on its Web site.

- In an April 2005 poll conducted by the Foundation for Biomedical Research, **76 percent** of respondents said that animal research has contributed either a great deal or a fair amount to advances in human health care of the past few decades.

- In a May 2007 Gallup poll of more than 1,000 Americans, **59 percent** said that they were in favor of medical testing on animals—a significant change from a poll conducted during the 1950s, when **84 percent** of respondents approved the use of animals in medical research and teaching.

Animal Rights Extremism

According to the Federal Bureau of Investigation (FBI), the greatest domestic terrorist threat in the United States is from environmental and animal rights extremists who have turned to threats, arson, vandalism, and explosives to get their message across. Between 1981 and 2005, these groups were allegedly responsible for more than 1,200 crimes and over $110 million in property damage. The following chart shows a breakdown of the various crimes that have been committed.

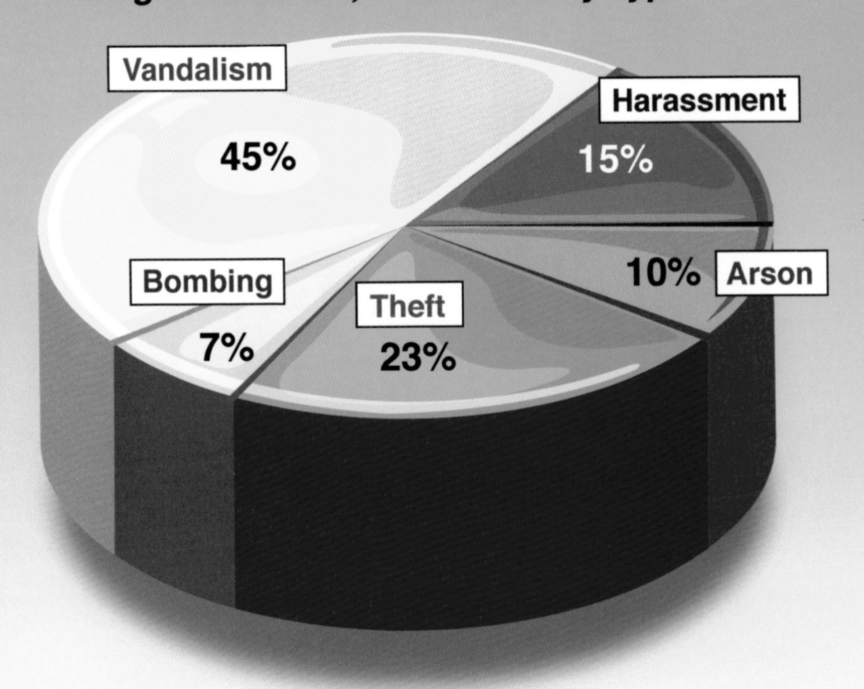

Source: Foundation for Biomedical Research, "Illegal Incidents Report: A 25 Year History of Illegal Activities by Eco and Animal Extremists," February 2006. www.fbresearch.org.

Changes in Animal Use and Testing, 1973–2005

While the use of most animals for experimentation and product testing has decreased over the past several decades, primate use has increased slightly.

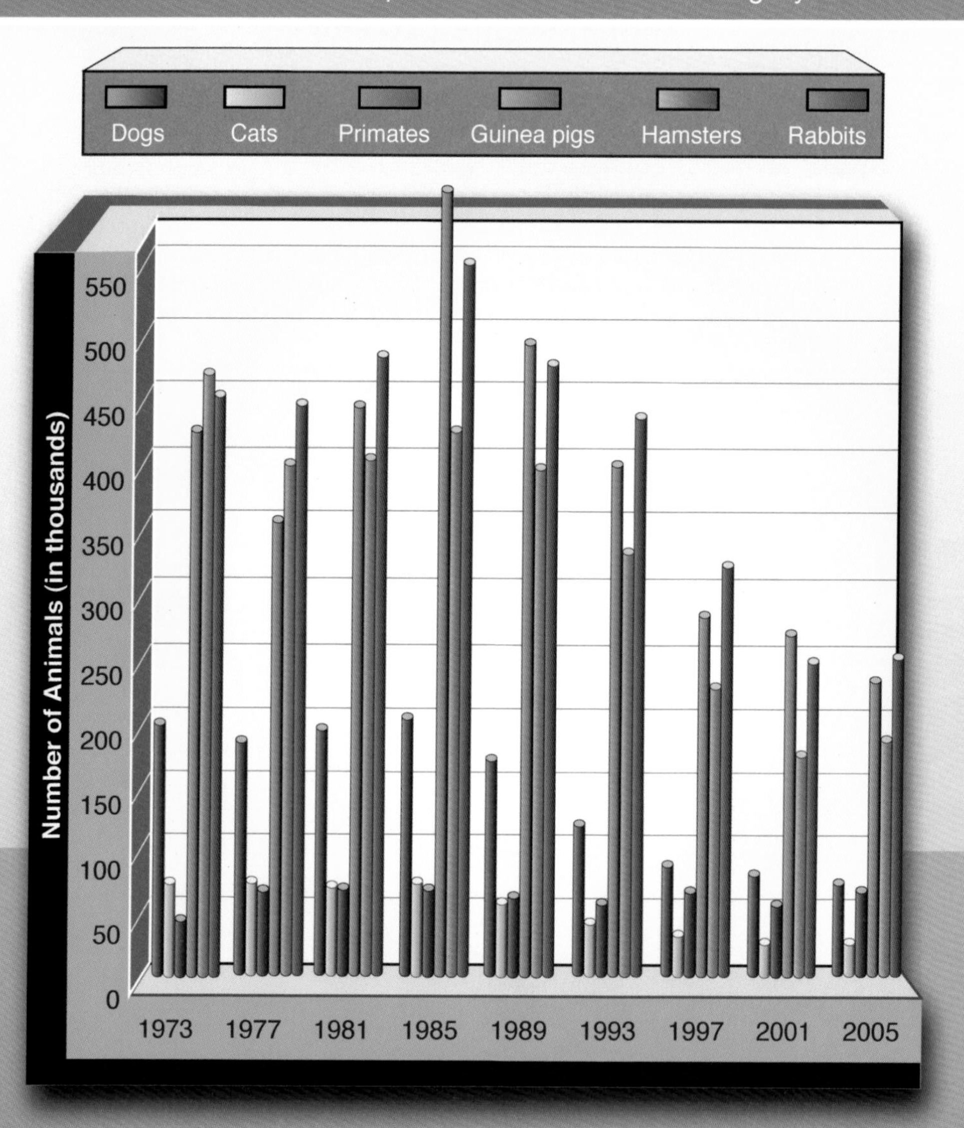

Source: United States Department of Agriculture, Animal, and Plant Health Inspection Service, "FY 2005 AWA Inspections," www.aphis.usda.gov.

Animal Research and Public Opinion

According to an April 2005 poll by Peter D. Hart Research Associates, a large majority of Americans support humane and responsible animal research for achieving medical progress, with 76 percent of participants saying that animal research has contributed to advances in human health care either a "great deal" or a "fair amount."

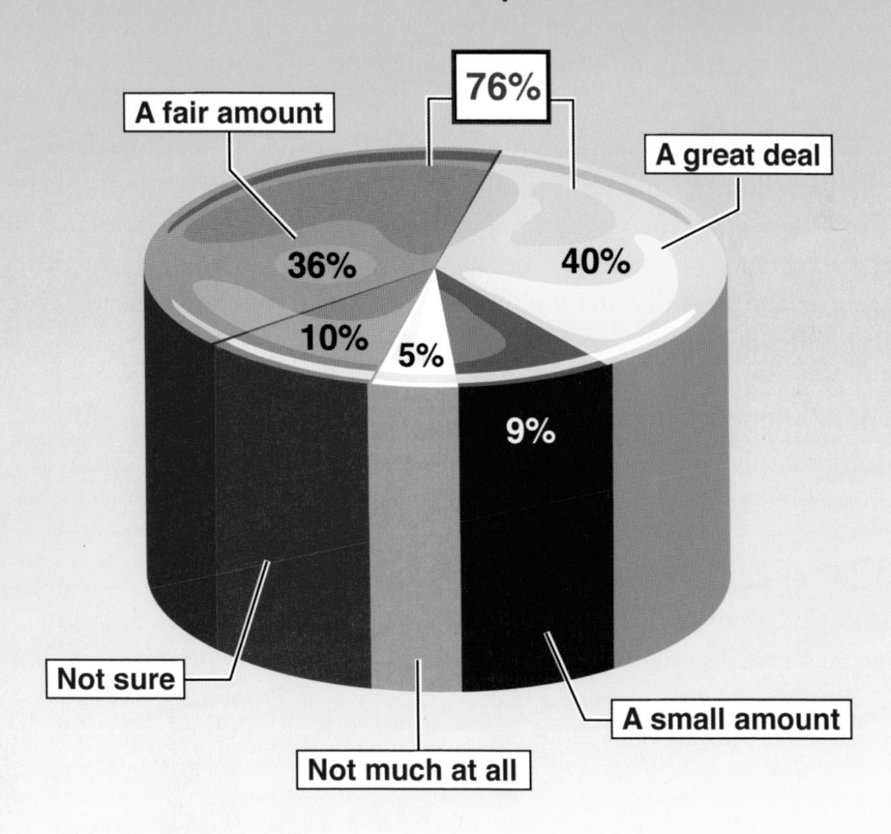

Source: Foundation for Biomedical Research, "Polls Show Majority of Americans Support Animal Research," June 3 2005. www.fbresearch.org.

Key People and Advocacy Groups

American Association for the Advancement of Science (AAAS): AAAS is an organization that is dedicated to advancing a worldwide understanding of science, and which supports the use of animals in scientific research.

Henry Bergh: In 1866 Bergh founded the American Society for the Prevention of Cruelty to Animals, which was the first animal rights group in the United States.

Claude Bernard: A French physiologist, Bernard published *Introduction to the Study of Experimental Medicine* in 1865 and gained much support from scientists for his stance on the merits of vivisection.

Charles Darwin: Darwin was a British scientist who challenged religious teachings of the 19th century with his book on evolution called *The Origin of Species by Means of Natural Selection.* Darwin influenced the acceptance of vivisection because he was a supporter of it, but his theories also increased the number of people who were opposed to animal experimentation.

Foundation for Biomedical Research (FBR): FBR is a powerful organization that advocates and promotes the necessity of humane animal research in an effort to improve human and animal health.

Galen of Pergamum: A Roman physician who lived during the second century, Galen wanted to learn more about the human body by studying anatomy. Through dissection of dead animals and experiments with live animals, he demonstrated that veins carry blood, not air, as scientists previously believed.

Jane Goodall: Goodall is famous throughout the world for her research on chimpanzees in Africa and for her compassion for primates. She has been awarded many prestigious awards and medals for her work, including recognition by the United Nations as a "Messenger of Peace" and being named by Great Britain's Queen Elizabeth II as a Dame of the British Empire, the equivalent of a knighthood.

William Harvey: Harvey was a prominent English physician who conducted experiments on animals in the 1600s in order to better understand how blood circulated throughout their bodies.

Humane Society of the United States: The largest and most powerful animal protection organization in the world, the Humane Society is against what it calls "unacceptable research and testing practices," such as research that involves primates.

Crawford Long and William Morton: Long and Morton accomplished a major medical breakthrough with their development of ether, an anesthetic that was used on animals to help alleviate their pain and suffering during research experiments.

Louis Pasteur: Pasteur was a noted French chemist who developed a vaccine for rabies in 1885 by using material from the spinal cords of rabbits he had infected with the virus.

People for the Ethical Treatment of Animals (PETA): Founded by Alex Pacheco and Ingrid Newkirk in 1980, PETA now has more than 1.6 million members and is the largest animal rights organization in the world.

Jonas Salk and Albert Sabin: Salk and Sabin were American scientists who each developed a vaccine against poliomyelitis following extensive research with primates. Salk's was made of inactive, or killed, poliovirus, and it was given as an injection. Sabin used live poliovirus in his vaccine, and it was designed to be taken orally.

Gary Yourofsky: An ex-employee of PETA, Yourofsky is active in the extreme animal rights group Animal Liberation Front (ALF), which is officially classified by the FBI as a terrorist organization. In 1996 Yourofsky founded a nonprofit group called ADAPTT (Animals Deserve Absolute Protection Today and Tomorrow).

Chronology

Second century
Galen of Pergamum, a physician from Rome, becomes one of the first scientists to experiment on animals and relate his findings to human beings.

1275
Ether is discovered by Spanish chemist Raymundus Lullius, who names it "sweet vitriol," but its anesthetic properties are not yet known.

1600s
English physician William Harvey discovers that the heart causes blood to travel in a circular pattern throughout the body, and in 1628 he publishes his findings in *On the Movement of the Heart and Blood in Animals.*

1822
The first anticruelty act is proposed in Great Britain by Richard Martin, giving cattle, horses, and sheep protection through the British parliament.

1824
Public dissections performed in London by French physiologist François Magendie lead to public outrage; that same year, Martin and 21 others form the first Society for the Prevention of Cruelty to Animals (SPCA).

1840
The SPCA's work is held in such high regard that England's Queen Victoria gives permission for the group to be called the Royal Society for the Prevention of Cruelty to Animals.

1842
The first surgery using ether as an anesthetic is performed by physician Crawford Long and William Morton.

1859
Charles Darwin challenges accepted religious beliefs and writes *The Origin of Species by Means of Natural Selection*, in which he presents the theory of evolution.

1866
New York native Henry Bergh founds the American Society for the Prevention of Cruelty to Animals.

1873
The American Anti-Vivisection Society is established.

1876
The British Cruelty to Animals Act is passed, becoming the first law that regulates animal experimentation.

1885
French chemist Louis Pasteur makes medical history by creating a vaccine for rabies.

1920s
Through experiments with dogs, Canadian scientists Frederick Banting and John Macleod discover insulin to control diabetes.

1951
Christine Stevens founds the Animal Welfare Institute in the United States.

1952
Jonas Salk develops the first vaccine for poliomyelitis.

1954
The Humane Society of the United States is established.

1958
Albert Sabin develops a live-virus polio vaccine to be administered orally.

1300 1600 1850 1900 1950

1959
British scientists William Russell and Rex Burch introduce the three Rs of animal experimentation: replacement, reduction, and refinement.

1963
The National Institutes of Health publishes the first guidelines for the care and use of lab animals in *The Guide for the Care and Use of Laboratory Animals.*

1966
The Animal Welfare Act becomes law in the United States.

1970
The AWA is amended to protect all warm-blooded animals used in research, experimentation, and testing, including requiring pain relief.

1971
The AWA is amended to state that rats, mice, and birds are not protected species.

1975
Philosopher Peter Singer publishes a book called *Animal Liberation: A New Ethics for Our Treatment of Animals*, which serves as a catalyst for an organized animal rights movement in the United States.

1980
Ingrid Newkirk and Alex Pacheco form the animal rights group People for the Ethical Treatment of Animals (PETA).

1985
Congress again amends the AWA, mandating that pain and suffering of animals be minimized with anesthesia and other drugs. Other revisions include exercise requirements for dogs and guidelines for ensuring the psychological well-being of primates.

1987
Jenifer Graham, a student from California, sues her high school for the right to refrain from participating in classroom dissection.

1989
Animal rights groups petition the USDA to reverse its 1971 regulation, thereby including rats, mice, and birds in its coverage.

1990
The USDA denies the animal rights groups' petition, and the Animal Legal Defense Fund and the Humane Society of the United States sue the USDA, prompting the USDA to extend AWA coverage to horses and other farm animals used in research.

1992
The U.S. District Court judge Charles R. Richey grants a summary judgment, ordering the USDA to reconsider coverage of birds, mice, and rats, but his decision is later overturned by the U.S. Court of Appeals.

1993
The first World Congress on Alternatives is held in the United States.

2000
The USDA agrees to modify its definition of the word *animal* to include birds, rats, and mice, thereby granting protection under the AWA.

2002
The AWA is amended with passage of the Farm Security and Rural Investment Act, which again excludes rats, mice, and birds from protection.

2006
The Animal Enterprise Terrorism Act is passed, which provides federal authorities with more ability to prevent, investigate, and prosecute violent crimes by extremist organizations.

2007
New Jersey becomes the ninth state in America to pass a Choice in Dissection law, which gives secondary school students the right to choose alternative methods over traditional animal dissection.

1960 1970 1980 1990 2000

Related Organizations

American Anti-Vivisection Society (AAVS)

801 Old York Rd., Suite 204

Jenkintown, PA 19046

phone: (215) 887-0816

fax: (215) 887-2088

e-mail: aavs@aavs.org

Web site: www.aavs.org

The American Anti-Vivisection Society states that its mission is to "unequivocally oppose and work to end experiments on animals and oppose all forms of cruelty to animals." Founded in 1883, the organization seeks to end the use of animals in research, testing, and education. AAVS publications include the quarterly *AV Magazine* and *Activate for Animals*, a bimonthly newsletter.

American Association for Laboratory Animal Science (AALAS)

9190 Crestwyn Hills Dr.

Memphis, TN 38125-8538

phone: (901) 754-8620

fax: (901) 753-0046

e-mail: info@aalas.org

Web site: www.aalas.org

The American Association for Laboratory Animal Science is an organization devoted to the humane care and use of laboratory animals in scientific research. It provides a forum for members to exchange information, holds an annual meeting for educational presentations, and publishes a number of journals and newsletters targeted at the laboratory science community, including *About Comparative Medicine* and *TechTalk*.

American Association for the Advancement of Science (AAAS)

1200 New York Ave. NW

Washington, DC 20005

phone: (202) 326-6400

fax: (202) 371-9526

e-mail: webmaster@aaas.org

Web site: www.aaas.org

The AAAS is an international organization that is dedicated "to advancing science by serving as an educator, leader, spokesperson, and professional association," and which supports animal research. In addition to organizing ongoing activities that advance an understanding of science, AAAS publishes the journal *Science* as well as other scientific newsletters, books, and reports.

Americans for Medical Progress Educational Foundation

908 King St., Suite 301

Alexandria, VA 22314

phone: (703) 836-9595

fax: (703) 836-9594

e-mail: info@amprogress.org

Web site: www.amprogress.org

The mission of the Americans for Medical Progress Educational Foundation is to enhance the public's understanding of and support for the necessity of using animals in medicine. The organization holds regular forums and campaigns to spread its message and educate the public about the value of animal research; and it publishes a number of documents including *Why Animal Research?*, *The Challenges of Animal Research*, and *Animal Research Saves Human and Animal Lives.*

American Veterinary Medical Association (AVMA)

1931 N. Meacham Rd., Suite 100

Schaumburg, IL 60173-4360

phone: (847) 925-8070

fax: (847) 925-1329

e-mail: avmainfo@avma.org

Web site: www.avma.org

The American Veterinary Medical Association, one of the largest and oldest nonprofit associations for veterinarians in the world, represents more than 75,000 veterinarians and calls itself "a collective voice for its membership and for the profession." The AVMA works closely with other organizations such as the Centers for Disease Control and Prevention and the National Academy of Sciences to accomplish its work in protecting animal and human health; and it makes available a variety of publications on animal health topics designed for veterinarians, government and regulatory agencies, and the public.

British Union for the Abolition of Vivisection

16a Crane Grove

London N7 8NN, United Kingdom

phone: +44 (0) 20 7700 4888

fax: +44 (0) 20 7700 0252

e-mail: info@buav.org

Web site: www.buav.org

The British Union for the Abolition of Vivisection describes its vision as, "To create a world where nobody wants or believes we need to experiment on animals." The organization works to seek changes in policies, legislation, and practices by educating the public and influencing lawmakers and decision makers to stop using animal research. It publishes a number of reports and other documents including *Next of Kin: A Report on the Use of Primates in Experiments.*

Centers for Disease Control and Prevention (CDC)

1600 Clifton Rd.

Atlanta, GA 30333

phone: (404) 639-3534 or toll-free (800) 311-3435

fax: (800) 553-6323

e-mail: inquiry@cdc.gov

Web site: www.cdc.gov

The CDC, which is part of the U.S. Department of Health and Human Services, is charged with the mission of promoting health and quality of life by controlling disease, injury, and disability. Its vision is "Healthy People in a Healthy World Through Prevention." The organization advocates for humane animal experimentation and research and provides data and statistical information, educational publications about health- and disease-related issues, and animal research documents.

Foundation for Biomedical Research

818 Connecticut Ave. NW, Suite 900
Washington, DC 20006
phone: (202) 457-0654
fax: (202) 457-0659
e-mail: info@fbresearch.org
Web site: www.fbresearch.org

The Foundation for Biomedical Research is dedicated to improving human and animal health by promoting the need and value of humane animal research and by refuting the arguments of animal rights organizations. The organization works to educate the public and media about the value of animal research, and its Web site offers position papers, *Illegal Incident* reports of criminal activities, and various other publications.

Humane Society of the United States

2100 L St. NW
Washington, DC 20037
phone: (202) 452-1100
fax: (202) 653-5760
e-mail: info@hsus.org
Web site: www.hsus.org and www.humaneteen.org

The Humane Society, the world's largest and most powerful animal protection organization, speaks out against what it calls "unacceptable research and testing practices," such as research that involves great apes. It works to reduce suffering of animals by advocating for public policies that protect them, investigates animal cruelty allegations, and educates the public about animal welfare issues. A number of publications can be downloaded from the Web site related to general animal research, species used in animal research, pain and distress suffered by laboratory animals, and animals in education.

National Anti-Vivisection Society

53 West Jackson Blvd., Suite 1552
Chicago, IL 60604
phone: (312) 427-6065 or toll-free (800) 888-6287
fax: (312) 427-6524
e-mail: feedback@navs.org
Web site: www.navs.org

The National Anti-Vivisection Society's tagline, "Advancing science without harming animals," encapsulates its mission, which is to abolish the use of animals in research, education, and product testing. The organization attempts to broaden the public's awareness of animal abuse and suffering through education programs, scientific research, and a variety of publications.

People for the Ethical Treatment of Animals (PETA)

501 Front St.
Norfolk, VA 23510
phone: (757) 622-7382
fax: (757) 628-0789
e-mail: info@peta.org
Web site: www.peta.org

With more than 1.6 million members worldwide, PETA is the largest animal rights organization in the world. It focuses on four separate areas where it believes the greatest amount of animal suffering occurs, two of which are factory farms and laboratory research. PETA's activities include education, investigations of animal cruelty, research, legislation, public service announcements, and public protests. Its Web site makes a number of publications available including fact sheets, photographs, and literature.

Physicians Committee for Responsible Medicine (PCRM)

5100 Wisconsin Ave. NW, Suite 400
Washington, DC 20016
phone: (202) 686-2210
fax: (202) 686-2216
e-mail: pcrm@pcrm.org
Web site: www.pcrm.org

Physicians Committee for Responsible Medicine promotes preventive medicine through innovative programs related to meat-free diets and nutrition plans, disease prevention and treatment, and overall health management. The organization advocates alternatives to animal research, opposes what it perceives to be unethical experiments in animals, and promotes nonanimal research in medical education. PCRM publishes a monthly online newsletter, an annual report, videos, and makes available a variety of research materials.

For Further Research

Books

Diane L. Beers, *For the Prevention of Cruelty: The History and Legacy of Animal Rights Activism in the United States.* Athens, OH: Swallow Press/ Ohio University Press, 2006.

Nancy Day, *Animal Experimentation: Cruelty or Science?* Berkeley Heights, NJ: Enslow, 2000.

C. Ray Greek and Jean Swingle Greek, *Sacred Cows and Golden Geese.* New York: Continuum, 2000.

Anita Guerrini, *Experimenting with Humans and Animals: From Galen to Animal Rights.* Baltimore: Johns Hopkins University Press, 2003.

David M. Haugen, *Animal Experimentation.* Detroit, MI: Greenhaven, 2007.

R.E. Hester and R.M. Harrison, *Alternatives to Animal Testing.* Cambridge, UK: Royal Society of Chemistry, 2006.

Karen Judson, *Open for Debate: Animal Testing.* New York: Marshall Cavendish, 2006.

Ellen Frankel Paul and Jeffrey Paul, *Why Animal Experimentation Matters.* New Brunswick, NJ: Transaction, 2001.

Bernard E. Rollin, *Animal Rights and Human Morality.* Amherst, NY: Prometheus, 2006.

Deborah Rudacille, *The Scalpel and the Butterfly: The War Between Animal Research and Animal Protection.* New York: Farrar, Straus, and Giroux, 2000.

Steven M. Wise, *Drawing the Line: Science and the Case for Animal Rights.* Cambridge, MA: Perseus, 2002.

Periodicals

Kathy Archibald, "Animal Testing: Science or Fiction?" *Ecologist Magazine,* May 2005.

Jessica Best, "Animal Testing Legislation Passes New York Assembly," *WWD,* June 15, 2007.

Greg Bolt, "Vegetables Shield Mice from Cancer in OSU Study," *Register-Guard* (Eugene, OR), December 19, 2006.

B. Bower, "Ape Aid: Chimps Share Altruistic Capacity with People," *Science News*, June 30, 2007.

Jerome Burne, "Animal Testing Is a Disaster," *Guardian*, May 24, 2001.

Sharon Pian Chan, "UW Professor Barred from Animal Testing," *Seattle Times*, November 19, 2003.

Andy Coghlan, "Report Claims Experiments on Monkeys Are Vital," *New Scientist*, June 2, 2006.

Clive Cookson, "Animal Testing on the Rise as More GM Rodents Used," *Financial Times*, July 25, 2006.

———, "Scientists Launch Public Campaign on Animal Tests," *Financial Times*, June 3, 2006.

Keay Davidson, "Fruit Fly Guinea Pigs Heading into Space; NASA Experiment on Shuttle Will Test Immune System," *San Francisco Chronicle*, June 30, 2006.

Stuart W.G. Derbyshire, "Time to Abandon the Three Rs," *Scientist*, February 2006.

Steve Doughty, "Why We Must Use Monkeys for Research, by Scientists," *Daily Mail* (London), June 3, 2006.

Felice J. Freyer, "Research Center Brings Scientists, Technology Together," *Providence Journal* (Providence, RI), October 8, 2006.

Brenda Goodman, "2 Soft Drink Giants to Curb Animal Testing," *International Herald Tribune*, June 1, 2007.

Frederick K. Goodwin and Adrian R. Morrison, "Science and Self-Doubt: Why Animal Researchers Must Remember That Human Beings Are Special," *Reason*, October 2000.

Jehangir S. Pocha, "Animal Tester," *Forbes Global*, October 30, 2006.

———, "Comparative Advantage (Animal Testing)," *Forbes*, November 13, 2006.

James Randerson, "Actors Make the Fur Fly," *Guardian*, March 14, 2006.

Anne Riley-Katz, "Protestors Force Juice Maker to End Testing on Animals," *Los Angeles Business Journal*, January 22, 2007.

Chris Shaw, "We Must Not Close the Door on Victims of Terrible Diseases," *Daily Telegraph* (London), May 18, 2007.

Jo Tanner, "Standing Up for Animal Research," *Chemistry and Industry*, May 1, 2006.

Sarah Treffinger, "Doctor in Dog Demo Barred from Doing Animal Research," *Plain Dealer* (Cleveland, OH), February 1, 2007.

David Wahlberg, "Scientists Report New Findings on HIV Transmission," *Wisconsin State Journal*, December 3, 2006.

Adam Wishart, "What Felix the Monkey Taught Me About Animal Research," *Mail* (London), November 26, 2006.

Internet Sources

Altweb: Alternatives to Animal Testing, Johns Hopkins University Center for Alternatives to Animal Testing, 2007. http://altweb.jhsph.edu/about.htm.

Christopher Anderegg, Kathy Archibald, Jarrod Bailey, Murry J. Cohen, Stephen R. Kaufman, and John J. Pippin, "A Critical Look at Animal Experimentation," Medical Research Modernization Committee, 2006. www.mrmcmed.org/Critical_Look.pdf.

Alan M. Goldberg, "Animals and Alternatives: Societal Expectations and Scientific Need," Johns Hopkins University Center for Alternatives to Animal Testing, October 6, 2004. http://caat.jhsph.edu/publications/articles/FRAME2004.pdf.

Humane Society of the United States, *Science and Conscience: The Animal Experimentation Controversy, a Study and Activity Guide for High-School Students and Their Teachers*, 2004. www.humaneteen.org/files/pdf/Science_and_Conscience.pdf.

Steven Milloy, "Laboratory Animal Farm," Fox News.com, April 4, 2001. www.foxnews.com/story/0,2933,2121,00.html.

Oregon Health & Science University, *Why Use Animals in Research?* http://onprc.ohsu.edu/emplibrary/animalsInResearch.pdf.

Crystal Schaeffer, "Allow Me to State the Obvious: Dogs Are Not Toasters," *AAVS Magazine*, Spring 2004. www.aavs.org/images/spring2004.pdf.

Source Notes

Overview

1. Deborah Rudacille, *The Scalpel and the Butterfly: The War Between Animal Research and Animal Protection.* New York: Farrar, Straus and Giroux, 2000, p. 25.
2. Michigan Society for Medical Research, "Fact vs. Myth." www.mismr.org.
3. National Anti-Vivisection Society, "Animals in Product Testing," www.navs.org.
4. Quoted in Stephen R. Kaufman, "Problems with the Draize Test," Americans for Medical Advancement. www.curedisease.com.
5. Quoted in Kathy Archibald, "Animal Testing: Science or Fiction?" *Ecologist Magazine,* May 2005. www.theecologist.co.uk.
6. Alex Pacheco, with Anna Francione, "The Silver Spring Monkeys," in Peter Singer, *In Defense of Animals.* New York: Basil Blackwell, 1985, pp. 135–47.
7. "The Humane Care and Treatment of Laboratory Animals," *NABR Issue,* 1999. www.nabr.org.

Do the Benefits of Animal Experimentation Justify Its Use?

8. Quoted in "Animal Research at CDC: Achieving a Delicate Balance," Department of Health and Human Services, Centers for Disease Control and Prevention, November 16, 2006. www.cdc.gov.
9. Quoted in Robert M. Baird and Stuart E. Rosenbaum, *Animal Experimentation: The Moral Issues.* Buffalo, NY: Prometheus, 1991, p. 20.
10. Quoted in Archibald, "Animal Testing: Science or Fiction?"
11. Quoted in Alex Epstein, "Animal Rights Movement Is Cruelty to Humans," *San Diego Business Journal,* August 22, 2005, p. 46.
12. Gen. 1:28, *The Living Bible,* Wheaton, IL: Tyndale House, 1971, p. 1.
13. Quoted in Francis Darwin, ed., *The Life and Letters of Charles Darwin.* New York: Basic Books, 1959, pp. 382–83.
14. Stuart Derbyshire, speech at the Edinburgh Book Festival, August 19, 2002. www.instituteofideas.com.
15. Claire Andre and Manuel Velasquez, "Of Cures and Creatures Great and Small," Markkula Center for Applied Ethics, Spring 1988. www.scu.edu.
16. Quoted in Andy Coghlan, "Report Claims Experiments on Monkeys Are Vital," *New Scientist,* June 2, 2006. www.newscientist.com.
17. Quoted in Gill Langley, "Next of Kin: A Report on the Use of Primates in Experiments," British Union for the Abolition of Vivisection, June 2006. www.buav.org.

Should Animal Experimentation Be Used in Education?

18. Jonathan Balcombe, *The Use of Animals in Higher Education,* Humane Society, 2000. www.hsus.org.
19. Quoted in Balcombe, *The Use of Animals in Higher Education.*
20. Quoted in Ruth Padawer, "Dissection Just Doesn't Cut It in Today's Science Classroom," *Record* (Bergen County, NJ) May 13, 2007. www.wormdigest.org.
21. Becky Baumgartner, "In Favor of Animal Dissections," Kenston High School, May 25, 2005. www.kenston.k12.oh.us.

22. Quoted in Padawer, "Dissection Just Doesn't Cut It in Today's Science Classroom."
23. Jane Goodall, "Foreword," in Balcombe, *The Use of Animals in Higher Education.*
24. Physicians Committee for Responsible Medicine, "PCRM's Campaign to End 'Cruelty 101,'" *PCRM Research.* www.pcrm.org.
25. *New York Times*, "Student Plans to File Lawsuit in a Dispute over Dissection," May 16, 1987. http://query.nytimes.com.

Are Animal Experiments Conducted Humanely?

26. Oregon Health & Science University, *Why Use Animals in Research?* http://onprc.ohsu.edu.
27. Alan M. Goldberg, "Animals and Alternatives: Societal Expectations and Scientific Need," Johns Hopkins University Center for Alternatives to Animal Testing, October 6, 2004. http://caat.jhsph.edu.
28. Bob Dole, letter to John McArdle, March 19, 2001, Society for Animal Protective Legislation. www.saplonline.org.
29. U.S. Department of Agriculture, complaint filed before the secretary of agriculture, August 27, 2004. www.idausa.org.
30. Southwest Foundation for Biomedical Research, "Animals in Research." www.sfbr.org.
31. Rachel Menge and Michelle Thew, "Product Testing in the United States: A Need for Change," *AV Magazine*, Summer 2005. www.aavs.org.
32. Humane Society of the United States, "Dark Side of Beauty: BOTOX Kills Animals." www.hsus.org.
33. Jonathan Balcombe, "Beyond Animal Research: Depressed Mice Study Is Inhumane and Clinically Irrelevant," *PCRM Research*, March 2006. www.pcrm.org.

Can Alternative Methods Eliminate the Need for Animal Experimentation?

34. Association of Veterinarians for Animal Rights, "Position Statements." www.avar.org.
35. Madhusree Mukerjee, "Trends in Animal Research," *Scientific American*, February 1997, pp. 86–93.
36. Nell Boyce, "TraumaMan Offers Lifelike Practice for Med Students," NPR, April 30, 2005. www.npr.org.
37. Quoted in AnimalLearn.org, *Educator Center: Educator Testimonials*, 2006. http://animalearn.org.
38. Baumgartner, "In Favor of Animal Dissections."
39. *CDC News*, "CDC's Animal Research Fact Sheet," November 16, 2006. www.cdc.gov.

List of Illustrations

Index

About the Author

Peggy J. Parks holds a bachelor of science degree from Aquinas College in Grand Rapids, Michigan, where she graduated magna cum laude. She is a freelance author who has written more than 60 nonfiction educational books for children and young adults. Parks lives in Muskegon, Michigan, a town that she says inspires her writing because of its location on the shores of Lake Michigan.